NEW DIRECTIONS IN UNDERSTANDING DEMENTIA AND ALZHEIMER'S DISEASE

ADVANCES IN EXPERIMENTAL MEDICINE AND BIOLOGY

Recent Volumes in this Series

A Continuation Order Plan is available for this series. A continuation order will bring delivery of each new volume immediately upon publication. Volumes are billed only upon actual shipment. For further information please contact the publisher.

NEW DIRECTIONS IN UNDERSTANDING DEMENTIA AND ALZHEIMER'S DISEASE

Edited by

Taher Zandi

SUNY at Plattsburgh
Northeastern New York Alzheimer's Disease Assistance Center
Plattsburgh, New York

and

Richard J. Ham

SUNY Health Science Center at Syracuse
Central New York Alzheimer's Disease Assistance Center
Syracuse, New York

PLENUM PRESS • NEW YORK AND LONDON

Library of Congress Cataloging in Publication Data

New directions in understanding dementia and Alzheimer's disease / edited by Taher
    Zandi and Richard J. Ham.
        p.   cm. — (Advances in experimental medicine and biology; v. 282)
    Proceedings of a conference held May 25-26, 1989, in Plattsburgh, N.Y., sponsored
by the Northeastern New York Alzheimer's Disease Assistance Center.
    Includes bibliographical references and index.
    ISBN-13:978-1-4612-7917-4   e-ISBN-13:978-1-4613-0665-8
    DOI: 10.1007/978-1-4613-0665-8
    1. Alzheimer's disease — Congresses. 2. Dementia — Congresses. 3. Alzheimer's
disease — Social aspects — Congresses. 4. Dementia — Social aspects — Congresses. I.
Zandi, Taher, 1952-   . II. Ham, Richard J. III. Series.
    [DNLM: 1. Alzheimer's Disease — congresses. 2. Dementia — congresses. 3. Family —
congresses. W1 AD559 v. 282 / WM 220 N5324 1989]
RC523.N5   1990
616.8'31 — dc20
DNLM/DLC   90-14297
for Library of Congress   CIP

Although dosages of medications are quoted, physicians and others
must check the manufacturers' directions and/or package insert
before prescribing.

Proceedings of a conference on New Directions: Understanding
Dementia and Alzheimer's Disease, held May 25–26, 1989,
in Plattsburgh, New York

ISBN-13:978-1-4612-7917-4

Dedicated to our families, and to the families and individuals
whose lives are affected by dementia

Introduction

The management of Alzheimer's Disease and the related dementias is one of the major challenges to health care professionals and American society-at-large for the coming decade and the coming millennium. The rapid growth of the over-eighty-five population, the group which, as recent studies have confirmed and as many of us clinicians have long suspected, has an even higher prevalence than previously quoted of dementing disorders, is the major cause of this. We are thus challenged by, as Bernard Issacs used to call it, "the survival of the unfittest," as well as the optimistic approach of "bringing life to years," as John F. Kennedy said.

The fact is that we, as a society, tend to confuse "treatment" and "cure" (and "prevention"). As the proceedings of the conference which this book represents emphasize, there is considerable work going on about the potential prevention of, or at least the reduction of, symptomatology in these illnesses by interventions genetically, chemically, and so forth. However, the more we find out, the more complicated it becomes, and the more heterogeneous Alzheimer's and the related disorders appear to be, not only in their manifestations (as clinicians have long recognized) but also in the individual initiating and underlying processes. For these reasons, absolute preventive techniques or the likelihood of an intervention which will reverse the process in a high proportion of patients, do not appear to be just around the corner. This must not in any way deter us from aggressively continuing to look at these factors. The presenters at our conference stimulate one to realize how much we must concentrate our efforts in research into all of these potential avenues.

However, the inability to prevent or cure should not deter us from "treating," and anyone who tells me that there is "no treatment" for Alzheimer's and the related disorders has got it all wrong: there is a great deal we can do now, as health professionals and informed caregivers, not only to relieve the suffering and burden of Alzheimer's, but even to actively promote health and function, despite the inexorable progression of the disease itself. Behavioral techniques, limited utilization of specific medications, environmental considerations, methods for preventing and handling crises and disturbances, for minimizing the disruption of the sleep disorders, catastrophic reactions and wandering, to avert aggressive behaviors--these are but some of the methods that we already have, but are as yet failing to implement for many patients and families afflicted with these problems.

Hence the significance of this conference and our book. The conference represented a coming together of scientists and researchers, clinicians of several health disciplines, informed caregivers and those who seek to educate them, and other interested individuals, including administrators and academicians, to focus on what we know, and what we can do right now, about these illnesses and the impact they are having. The conference represents a model of the ways in which, on a regional basis, this rich mixture of individuals must unite in their efforts to find the resources and the strengths, and to disseminate the necessary information, in order to make a real impact on Alzheimer's and the other dementias in our immediate future, as well as in the coming decades.

Richard J. Ham, M. D.
SUNY Distinguished Chair in Geriatric Medicine

March 1990

Preface

On May 25 and 26, 1989, a conference entitled New Directions: Understanding Dementia and Alzheimer's Disease was held in Plattsburgh, New York. This conference was sponsored by the Northeastern New York Alzheimer's Disease Assistance Center at the State University of New York campus in Plattsburgh. The conference brought together participants and presenters from across the State and beyond, based around the network of Alzheimer's Disease Assistance Centers, founded in 1988 by the New York State Legislature. It was a multi-disciplinary conference with presentations by practitioners, clinicians, epidemiologists, scientists, social service providers and caregivers.

This conference was organized in response to the abundance of recent demands regarding the multi-faceted aspects of dementia and the paucity of multi-disciplinary workshops and conferences.

At this conference, the presenters explored, synthesized, redefined and most importantly, reiterated their commitment to meeting the challenges and obstacles that are before them.

The conference addressed a variety of topics, including laboratory investigations of chromosome 21, diagnostic procedures, etiology of Alzheimer's Disease, reversible dementia, case management and the role of support groups for dementia caregivers.

This book is a fully edited version of the proceedings of this conference and is dedicated to dementia patients and their caregivers who suffer together in the seemingly endless battle against this disorder. This book, like other contributions in this field, reflects hope and optimism--hope for the day when we can conquer the pain, agony and fear of this silent epidemic in human society; optimism in that we have now a greater understanding of this disorder and current means to relieve some of the problems and even improve some aspects by careful management and a well-informed, educated approach by all involved.

Our understanding of this disorder has greatly increased over the past decade. We are far from a breakthrough in terms of prevention or cure of the disease itself, but we now are better at differentiating the symptoms that are caused by this disease itself, versus those that are induced by the nature of the environment that the disease created, by concurrent disorders, and by the social isolation of the Alzheimer's patient and the family caregiver. These distinctions enable us to implement treatment procedures more effectively.

The treatment procedures range from a complete case management of both patient and caregiver to drug treatment of the controllable symptoms. For example, tetrahydroaminoacridine (THA) is one of several drugs being tested as treatment for Alzheimer's patients. THA may help control memory loss in some patients with AD. The preliminary results from this study will be available soon.

Another recent research approach concerning Alzheimer's patients' brain functions has been geared towards brain-repair mechanism. Alzheimer's Disease anatomically causes extensive brain injury. Scientists are looking at the brain-cell repair and its elasticity analogous to the brain repair that occurs after a trauma or brain injury.

The research approach toward identifying the genetic markers in Alzheimer's Disease has also progressed in the past decade. Studies have shown that a small percentage of AD patients have shown an inherent pattern suggestive of autosomal dominant transmissions. More recently, scientists have identified a particular locus on chromosome 21 as being responsible for production of the abnormal amyloid protein.

Most significantly in terms of what we can do now, the scientific community working with AD patients has fully realized that AD patients are not the only victims of this disease. The caregiver's well being is being taken into consideration, and the caregiver's burden is receiving national and international attention. During 1989, nearly thirty-five percent of the articles appearing in gerontology related journals in the United States involved caregiver issues.

This book, among other topics, addresses the above issues. The topics are ordered in three sections of medical, cognitive, and social issues associated with Alzheimer's Disease and other forms of dementia as well as the caregiver's issues and problems.

Organization of the two-day conference that lead to the development of this book was made possible through the dedicated efforts of the staff of the Northeastern New York Alzheimer's Disease Assistance Center. We wish to express our thanks to all the people who were involved in these efforts. Special thanks go to Gloria Bushey, Linda Patnode and Martha Cashman for their enormous efforts in the technical preparation of this manuscript.

Taher Zandi, Ph.D.

Contents

Section I

A Medical Agenda: Etiology, Diagnosis and Treatment of Dementia

Chapter 1

ALZHEIMER'S DISEASE AND THE FAMILY: A CHALLENGE OF THE NEW

MILLENNIUM

Richard J. Ham, M.D.

SUNY Distinguished Chair in Geriatric Medicine
Professor of Medicine
State University of New York
Health Science Center at Syracuse
Syracuse, New York

When Alzheimer described his case--a two-page case report in a German medical journal in 1907 (1), he described progressive dementia in a relatively young woman, who died at age 55. Alzheimer had been intrigued by the case because she seemed to be suffering from a premature version of the senile dementia which, even then, was well recognized as a common illness among the very old. Perhaps even Alzheimer himself thought, as many have since, that dementia in the elderly was just an exaggeration of "normal aging." He was impressed enough by the unusually early onset of this unfortunate woman's illness that he applied the newly discovered silver stains to the autopsy material, and it was there he discovered the classic plaques and tangles which are still pathognomic of the disease that bears his name. It took many decades before clinicians as a group accepted the fact that the illness he had described is in fact the same illness which produces the majority of cases of progressive dementia in the elderly; the recognition that this very specific illness, with its characteristic histopathological findings underlying this alarmingly common illness of the very old, has occurred only over the past decade or so. Until quite recently, otherwise clinically accurate physicians accepted vague diagnostic labels like "cerebral atherosclerosis" or "organic brain syndrome" as being acceptable diagnostic descriptions for the chronic brain failure they noted in many of their older patients. The name "Alzheimer's" has now caught on. The giving of a name to an illness might not seem very important, but to family members who have lived through the frequently prolonged and painful early phases of Alzheimer's, telling themselves perhaps that mother could "pull herself together" and take control of her situation as she always did, or having it implied that it is "normal" (an implication that may be made by well meaning friends and even by the personal physician), all add to stress which ought to be resolved as soon as it can be by a reasonable physician making a suitably thorough investigation in order to establish a working diagnosis of "probable Alzheimer's." Appropriately, many physicians worry about premature labeling, but the giving of a name to this illness is important: it must not be given lightly, and a suitable workup and a hunt for the reversible elements (and the considerably more rare truly "reversible" causes) must be done before the label is too firmly applied. But applied it must be, for the diagnostic label gives the family the dignity and definity of an illness to work with, gives them a name they can look up in the telephone directory for help (the Alzheimer's Association), and a name they can look up in the library for information. It has taken the average lifetime of the average victim of Alzheimer's

New Directions in Understanding Dementia and Alzheimer's Disease,
Edited by T. Zandi and R. J. Ham, Plenum Press, New York, 1990

for us to realize that the disease he described is indeed the one which afflicts such a high proportion of the very old.

It is because increasing age is the primary risk factor for the development of Alzheimer's that we hear so much about it nowadays. The striking increase in the population of this country of those "oldest of the old" (the over eighty-five cohort), and the increasing realization that the prevalence of dementia, most of it caused by Alzheimer's, may be as high as 50 or more percent in the over eighty-five population, combine to make us keenly aware of the urgency of our mission as health professionals: a two-fold mission, first to seek to understand the etiology and thus eventually cure, relieve or prevent, but, meanwhile, to treat the symptoms, signs and dysfunction that the illness presents in the patients themselves, and in their family members.

The demographics of the "old, old" are quite dramatic: we are currently well into a fifteen-year period during which the over eighty-five population in this country is doubling in number. There are currently 2.7 million people over eighty-five in the United States; by 2050 (when I will be 104!), there will be 15 million people over eighty-five in the U.S.--at that time, one-half, at least, of all Americans will live to see their eighty-fifth birthday (2). This is an incredible gift of our present civilization and development (See Table 1 in Chapter 9 of this book).

The bleak view that this demographic change could represent only the "suvival of the unfittest" is nurtured by the high incidence of Alzheimer's. It is also nurtured by the high incidence of other problems--for example, hip fracture (3). (About one-third of women in their late eighties will fracture a hip, one-sixth of men will, and a high proportion will die, or at least be chronically disabled, as a result). Now we are beginning to see ways in which hip fracture incidence may be reduced: better knowledge about falling and gait is one aspect, but better knowledge about osteoporosis, and its prevention by action earlier in life has quite changed our thinking about it. As you shall see later in these proceedings, if and when we can become microscopic biochemical engineers in repairing our DNA, probably somewhere on the 21st chromosome (4), then we may be talking about preventive measures in early life for Alzheimer's as well.

But clearly it will probably be a long time before this comes to pass. The more Alzheimer's is investigated and the more we find out, the more we realize its heterogeneous nature and multiple etiology--it may well represent one end point from several different diseases or factors; the identified neurophysiological and biochemical changes of Alzheimer's are multiplying: the thought that doing something about the acetylcholine lack (in the way that L-dopa provides a temporary replacement therapy of sorts in Parkinson's) might be "the answer" has been countered by the increasing realization that acetylcholine lack is but one of multiple nuerochemical deteriorations in the Alzheimer's patient. For now, our challenge is a challenge not of cure but of treatment. And, in this keynote presentation, it is the eminent treatability of Alzheimer's and the related dementias that I wish to emphasize. It is quite wrong to think of Alzheimer's as "untreatable." It is certainly incurable, but each and every manifestation (and, as we become more experienced, we become more specific at identifying the individual behavioral and symptomatic manifestations) can and should have a treatment plan established: the reason that my title is "Alzheimer's and the Family" is that much of the management and treatment relies on a family member, or some other caregiver, generally not a health professional, generally the person who is there for most of the "thirty-six hour day" that is involved in caring for these patients (5). Good management depends on that person gaining the strength, knowledge and skill necessary to anticipate, handle and minimize each effect. We should regard the informed caregiver as the therapeutic "right arm" of the health professional--the means whereby the treatment and management plan of the established dementia patient is implemented.

RECOGNITION AND DIAGNOSIS

Williamson's classic work concerns what we have come to call the "icebergs" of geriatric medicine (6)--the many common and disabling conditions of elders which, for one reason or another, are demonstrably not known about by the personal physician, nor frequently by the family members--problems such as dysmobility, incontinence, hearing loss, depression and, of course, dementia. The trouble, disruption and stress which can be caused during the painful early phases where the patient has insight, where the family may feel that the apparent personality change is within his/her control, is a phase of great family disruption. I can well remember, in my own family's case, my grandmother's previously dominating and often acerbic character becoming worse and worse, with we grandchildren finding her more and more intolerable, when in fact she was becoming ill, not just "awkward." No one ever properly diagnosed my grandmother's illness, although it is obvious in retrospect what it was. The result was that right up to the end unreasonable demands were being made of her in her matriarchal role, and no consistent effort was being made to minimize the frequently embarrassing effects of her mental confusion and memory loss on herself and her life. I was recently involved in another tragic case, of a woman with insight but evident cognitive loss who really wanted to know what the illness was, so I told her. She had soon forgotten what I said, but her husband didn't: he was furious. A dominating man who cared a great deal about what the community and neighbors thought, he was himself terminally ill, and disallowed any mention of the word, or any specific plan to help his wife, whom he frequently challenged and tested in his desperate bids to prove to himself that she was still the same person, and not unwell at all. He was dying and is now dead. Meanwhile his wife has deteriorated; you can say it was a mercy that she did not realize what was going on even on the day of his funeral, but for the family it has been a double loss. "If only Dad had come to terms with the illness, we could have made more efforts and been able to enjoy her while there was still something left." How sad to miss out on the last available part of someone's actual mental life and personality--"the person," who, as so many Alzheimer's caregivers say, dies long before physical death.

So we must all be alert to early recognition of this illness (and the other dementias), and exhort our professional colleagues into recognizing just how frequently apparent but persistent personality or behavioral change, loss of function or apparent self-neglect, or sometimes apparently transient disorientation in extraordinary circumstances (like on a vacation or at a family function) are, in fact, the outward signs of the progressive cognitive loss of a dementia. It is at this phase of early (almost "prodromal") symptoms that it can be sometimes helpful to apply formal mental status testing as a screening tool. Such an instrument as the Folstein (7) can reveal cognitive loss in those situations where ordinary social conversation, with short exchanges of familiar phrases, would not reveal the deficit (as it has often not been revealed to family members). Many of us have seen the examples of family members truly appalled to realize just how lost mentally their relative has become, despite the social skills. Remember that such an authoritative figure as Katzman (8) has opined that as many as 10% of Alzheimer's patients do suffer a truly prodromal phase with frequently some only slowly progressive single cognitive deficit continuing for several years, even before the characteristic more widespread cognitive losses start to increase.

These early phases of Alzheimer's can be quite a problem to the clinician and it often would be inappropriate to diagnose too emphatically at this point. As with many illnesses in elders, whereas the best attempts possible should be made to produce an accurate diagnosis, there may well need to be a period of toleration of ambiguity about the diagnosis, and again the use of formal mental status testing can be valuable. Appropriately applied in an early case in which there is suspicion of the diagnosis, it can provide confirmatory evidence especially when used to verify decrements as they develop progressively over time (as they inevitably will if the diagnosis is Alzheimer's).

It must be emphasized that, whereas screening instruments such as Folstein's Mini-Mental State Examination are useful, they are sometimes not particularly sensitive. A very intelligent person may score amazingly well, even in the presence of highly suggestive symptoms, and even when, as becomes clear in retrospect as the disease progresses, they were genuinely in the early phases of progressive dementia. Thus the Folstein, or similar instruments, never "rule out" Alzheimer's, but they can certainly "rule it in" at times. Such instruments should be much more widely used in primary care practice (9) and in evaluations by other health professionals where dementia is a possibility. The Folstein instrument is reproduced in Illustration 1. (See Chapter 8 for further details of the Mini-Mental State Examination.)

A hot issue is what investigations should be done, once broad cognitive impairment has been demonstrated, and once it is clear that several concurrent progressive cognitive losses are truly interfering with daily function (i.e., once the diagnostic criteria for dementia are fulfilled). The DSM III (10) criteria for dementia are reproduced in Table 1. By definition, for dementia to be diagnosed, other illnesses that might look like it have to be excluded.

The recent legislation from the Health Care Financing Administration (11), to do with accurate assessment of hospitalized patients heading towards nursing home placement, and the accurate assessment of patients said to be suffering from dementia and thus placed in nursing homes, provides a particularly timely focus on this issue. The well established mnemonic DEMENTIA (Table 2) (12) is helpful. Many series, particularly Larson et al. (13), have clarified that the incidence of factors that could conceivably, if treated, "reverse" and therefore "cure" the dementia, is very much higher than the actual incidence of truly "reversible dementia" (14,15). (See Chapter 4). In other words, a high proportion of the patients with serious thyroid disorders in need of treatment will have a concurrent progressive dementia anyway, and, whereas the thyroid condition must of course be treated, the dementia will progress unabated. This is no reason for negativity, since the natural history of dementias, with a duration of anything from three to fifteen or more years from early symptoms to death, makes it crucial to look for all contributory health factors. Whereas their treatment might not reverse the dementia, their treatment will likely ameliorate the symptoms of the dementia. The recommended minimal workup for a patient shown to have dementia is summarized in Table 3. The history is the most important part of this, and in obtaining it, since the caregiver must give much of the information for speed and reliability, the caregiver is appropriately being simultaneously assessed. This is vital, since the caregiver, and all the environmental factors around the patient, are at least as important as the active symptoms in terms of the real prognosis for future management and dependency.

Other presenters in these proceedings will be reviewing the investigations in detail. Suffice to say, in summary, that it should always be borne in mind that the effects of medication and thyroid disorders dominate as the two truly "reversible" types of dementia, providing one excludes the high incidence of depression in elders in which cognitive impairment is so dominant that it "looks like dementia." More about this in a moment.

It should also be emphasized that, although the new federal regulations seem to imply that brain imaging should always be done, most clinicians (and the Seattle series (13) provide us some evidence to support this) believe that the return on brain imaging is very low indeed, negligible in fact, in those patients who have clearly had symptoms for more than three years, with a consistent history of progression, no abnormal neurological symptoms or signs, a slow onset, and no recent abrupt deterioration or history of potential head injury. It would not be desirable for brain imaging to become a requirement for the diagnosis in such late cases with all the typical symptoms. It would be particularly inappropriate to ship a case like that described from a remote area, where brain imaging was not available, just for this study. It is also technically difficult to get some demented patients to lie still long enough, particularly for the MRI. The incidence of dementia is so high, that unnecessary

TABLE 1

Diagnostic Criteria for Dementia

Loss of intellectual abilities severe enough to interfere with social or occupational functioning
Memory Impairment
At least one of the following:
    Impairment of abstract thinking manifested by concrete interpretation of proverbs, inability to find similarities or differences between related words, difficulty defining words and concepts or other similar problems
    Impaired judgement
    Other disturbances of higher cortical function, such as aphasia (disorder of language due to brain dysfunction), apraxia (inability to carry out motor function), agnosia (failure to identify or recognize objects despite intact sensory function), "constructional difficulty" (e.g., inability to copy three-dimensional figures, assemble blocks or arrange sticks in specific designs)
    Personality change
State of consciousness not clouded (does not meet the criteria for delirium [Table 1] or intoxication, although these may be superimposed)
Either of the following:
    Evidence from the history, physical examination or laboratory tests of a specific organic factor that is judged to be etiologically related to the disturbance
    In the absence of such evidence, an organic factor necessary for the development of the syndrome can be presumed if conditions other than organic mental disorders have been excluded and if the behavioral change represents cognitive impairment in a variety of areas

(From Diagnostic and statistical manual of mental disorders, 3rd ed. [DSM-III]. Washington, D.C.: American Psychiatric Association, (c) 1980. Reprinted with permission.)

nvestigation should be avoided. What one needs is the time to take a good history: he fact that time is the one thing for which Medicare and all the other insurances 'ail to pay is a separate challenge of geriatric medicine which I will not start into 1ow. Hopefully, "comprehensive geriatric assessment" (CGA) (16) will become ·ecognized as a reimbursable procedure (several of us in geriatrics are working hard)n this). This is a comprehensive assessment that involves more than just a physician's :ime, it involves other health professionals, particularly a social worker and perhaps a 1urse clinician, and often others working together and which is necessary for many :omplex geriatric situations (the patient with dementia is certainly a prime example). [t is no good recommending spending time, if one is not simultaneously working on)ur health care system to make it an economically viable thing for health professionals :o do.

TABLE 2

Reversible Causes of Apparent Dementia

D	-	Drugs
E	-	Emotional illness (including depression)
M	-	Metabolic/endocrine disorders
E	-	Eye/ear/environmental deficits
N	-	Nutritional/neurologic conditions
T	-	Tumors/trauma
I	-	Infection
A	-	Alcoholism/anemia/atherosclerosis

(Reprinted with permission from Ham, R. Geriatrics II AAFP Home Study Self-Assessment Monograph 96. AAFP, Kansas City, Mo., 1987, AAFP.)

TABLE 3

The "Workup" in Potential Dementia

---

1. A comprehensive history including a drug/medication review

2. A complete physical examination especially neurological

3. A standardized mental status screening exam

4. Diagnostic studies, including

   (i) all of the following:

     a) complete blood count
     b) electrolyte metabolic panel
     c) screening metabolic panel
     d) thyroid function tests
     e) Vitamin B-12 and folate levels
     f) tests for syphilis and, depending on history, for
       human immunodeficiency antibodies

   (ii) in most cases:

     computed tomography of the brain (with contrast) or magnetic
     resonance imaging. (Optional if all of the following apply: history
     more than three years, onset gradual, no early neurological signs or
     symptoms, no question of head trauma, smooth progression of
     symptoms, and dementia already at least moderate)

NOMENCLATURE

The NIMH has published a uniform set of diagnostic criteria, evolved by consensus, that clinicians and researchers should now use (17). For the practicing physician, the most definite diagnosis in most instances is "probable Alzheimer's Disease." This is a very honest terminology (we are not usually so honest about our diagnoses, where the probability rate is often much less than the 90-95% which many experts feel can be achieved from a decent clinical history and diagnostic workup as implied above). The phrase "possible Alzheimer's disease" is a useful adjunct, to express the situation where potentially causative factors, particularly something like alcoholism, are found, and yet the possibility is there of this being concurrent true Alzheimer's disease. (These criteria are tabulated in Table 4.)

DEMENTIA AND DEPRESSION

The interrelationship between depression and dementia is a fascinating one. Wells coined the phrase "pseudodementia" to express those depression patients in whom cognitive impairment is so marked that the patient appears to be suffering a dementia (18). He has written elegantly about differences including the way in which the "depressed" as opposed to the "demented," patient responds to questions (Table 5). It is possibly true that, as in the TV documentary Bette Davis starred in, elders have sometimes been labeled as "dementia" when in fact it was a depression. In that TV drama, Miss Davis' apparent "dementia" resolves when the depression lifts (depression being a self-limited disease in all patients, provided suicide does not intervene), but by that time she has been institutionalized and has lost her apartment and much of her independence. This absolute differentiation should not be difficult, although it is the experience of specialized Alzheimer's clinics to find patients who truly are only suffering a depression with marked cognitive impairment. It is an exciting find, of course, for the illness can be treated with an anti-depressant and frequently the

TABLE 4

Criteria for Clinical Diagnosis of Alzheimer's Disease

Criteria for clinical diagnosis of PROBABLE Alzheimer's Disease
Dementia on clinical examination, documented by mental status tests and confirmed
  by neuropsychologic tests
Deficits in two or more cognitive areas
Progressive worsening of memory and other cognitive functions
No disturbance of consciousness
Onset between the ages of 40 and 90, usually after age 65
Absence of systemic or brain disorders that in themselves could account for the
  progressive memory and cognitive deficits

Evidence to support diagnosis of PROBABLE Alzheimer's Disease
Progressive deterioration of specific cognitive functions (e.g., language [aphasia],
  motor skills [apraxia], perception [agnosia]
Impairment of activities of daily living and altered behavior patterns
Family history of similar disorders (especially if pathologically confirmed)
Normal lumbar puncture results
Normal or nonspecific (e.g., slow-wave) changes in EEG
Progressive cerebral atrophy on serial CT scans

*Features consistent with PROBABLE Alzheimer's Disease (after exclusion of other causes
  of dementia than Alzheimer's Disease)*
Plateaus in the progression of the illness
Associated symptoms: depression, insomnia, incontinence, delusions, illusions,
  hallucinations, catastrophic reactions (verbal, emotional, physical), sexual
  disorders, weight loss
Other neurologic abnormalities (especially in advanced cases), including increased
  muscle tone, myoclonus and gait disorders
Seizures (in advanced cases)
CT scan normal for age

*Features inconsistent with PROBABLE Alzheimer's Disease (that make the diagnosis
uncertain or unlikely)*
Sudden apoplectic onset
Focal neurologic findings, such as hemiparesis, sensory loss, visual field defects and
  early incoordination
Early seizures or gait disturbances

Criteria for clinical diagnosis of POSSIBLE Alzheimer's Disease
Dementia syndrome with variations in the onset, presentation or course, in the
  absence of other disorders that in themselves may cause dementia
Dementia associated with a second system or brain disorder that would be sufficient
  to produce dementia but which is not considered to be the prime cause
Single, gradually progressive and severe cognitive deficit in the absence of an
  identifiable cause (this criterion especially designed for use in research studies)

Criteria for diagnosis of DEFINITE Alzheimer's Disease
Criteria for probable Alzheimer's Disease plus histopathologic evidence from biopsy
  or autopsy

*Suggested subgroups (features which may differentiate types of probable Alzheimer's
  and may be useful in research)*
Familial
Starting before age 65
Associated with trisomy 21
Coexistent with Parkinson's Disease or other relevant condition

*(From McKhann, G., et al. Clinical diagnosis of Alzheimer's disease: report of the
NINCDS-ADRDA Work Group under the auspices of Department of Health and Human
Services Task Force on Alzheimer's Disease. Neurology, 1984:34 (7): 939:44.
Reprinted with permission. (c) 1984.)*

depression will be lifted, and the ultimate prognosis is good for many. But there is more to it than that. For years, experienced clinicians have been wondering whether such cognitively impaired depressions in some way foreshadow, or even predispose to, dementia itself. Certainly one sees elderly depressives with cognitive impairment as a major feature who never truly "recover"--they never quite become themselves again, following the episode of depression. Recent work demonstrating how much structural change there is (seen on nuclear magnetic resonance imaging) in these patients verifies the "organicity" of much elderly depression (15). Also, it is now becoming clear that, as with other chronic progressive illnesses, there is an incidence of depression concurrent with dementia--some of it caused by realization about the illness, some truly "coincidental," but some of it surely related to our increasing knowledge that some of the neurophysiological changes in the dementias have interesting chemical structure parallels with the structure of some of the effective anti-depressant medications. Thus these two common illnesses of elders--although dementia is still much more common than depression--occur together frequently and in an interlinked fashion. A particular message to all clinicians is to watch for the development of depression in the established dementia patient, for when it occurs, not only might the patient truly become despondent and apathetic, and much more dysfunctional, or even suicidal (remember that fluctuating organic impairment or awareness is an extra risk factor for suicide), but this increased impairment comes at just a time when the illness is particularly stressful, and it will break even further the patient's or the family's spirit and resolve. And it is often quite treatable. Anti-depressants can be difficult to handle in dementia patients, but frequently, with sensitive follow-up and reasonable counseling of the caregiving family, an anti-depressant can be safely titrated in the situation of a depression complicating a dementia, with considerable benefit. At our Alzheimer's clinic, we are tending to use desipramine where we want psychomotor activity to improve and fluoxetine where a more serotoninergic, less psychomotor-active medication is needed. The low anticholinergicity of these two medications (fluoxetine virtually none) makes them very advantageous compared to the traditional tricyclics. (See Chapter 6, Newhouse's discussion on effect of cholinergic drugs on dementia and depression.) Before I close on this subject, I must mention ECT, which can be incredibly valuable in these patients and can quickly resolve a depression which may not respond at all to any oral antidepressant. The memory loss associated with unilateral ECT to the non-dominant hemisphere is transient and is not an issue in these patients in the way in which many suspect. There is much muddled thinking about ECT--it is a lifesaving procedure which has not become unfashionable in many Western countries in the ways in which it has over here.

PRESENTATION OF DEMENTIA: SOME PEARLS

- Social function is often astonishingly preserved despite severe loss of cognitive ability.
- The first and most vital question is always "How long has this been going on?" If the presenting change in mental status does truly appear to have been sudden, then the workup is urgent: a mental status change caused by an incidental acute illness may well represent life-threatening disease, or at least something that will cause irreversible brain damage if not acted on quickly.
- Alzheimer's assessment means assessing the caregiver as well as the patient. No plans can be made for the patient unless someone can accurately describe the home circumstances and environment.

SYMPTOMS AND BEHAVIORS IN THE ALZHEIMER'S PATIENT

At this point we will review in approximatly the order in which they generally occur some of the "typical" and common symptoms that will occur in many Alzheimer's patients. As noted above, the implementation of preventive and treatment plans for these various behaviors and manifestations rests largely with the caregiver under the guidance and education of the health professionals involved in the case (5).

TABLE 5

Distinguishing Depression From Dementia

Dementia	Depression
Insidious onset	Abrupt onset
Long Duration	Short duration
No psychiatric history	Often previous psychiatric history (including undiagnosed depressive episodes)
Conceals disability (often unaware of memory loss)	Highlights disabilities (in particular complains of the memory loss)
"Near-miss" answers	"Don't know" answers
Day-to-day fluctuation in mood	Diurnal variation in mood, but mood generally more consistent
Stable cognitive loss	Fluctuating cognitive loss
Tries hard to perform but is unconcerned	Often does not try so hard but is more distressed by losses
Memory loss greatest for recent events	Equal memory loss for recent and remote events
Memory loss occurs first	Depressed mood (if present) occurs first
Associated with unsociability, uncooperativeness, hostility, emotional instability, confusion, disorientation and reduced alertness	Associated with depressed or anxious mood, sleep disturbance, appetite disturbance and suicidal thoughts

(Adapted from Wells, C.E. Pseudodementia. Am J Psychiatry. 1979:36(7) 895-900. Reprinted with permission.)

Memory Loss

This is often the first symptom noted. Simple memory aid devices can be very helpful--just labeling, coding the telephone, reminder calls from the caregiver, etc. It is important that the family realize that "forcing" will not induce memory. A positive thing to emphasize early is that long term memory is relatively preserved. Long term memory can thus be enjoyed, even when short term memory is poor. In fact, family members need to be encouraged to take the time to collect together albums of old photographs, perhaps even records of appropriate music or tapes of TV shows, things from the past that can be enjoyed, that will help give structure and a sense of relevance to a person who is losing his/her sense of reality. In some cases, a book of old photographs can be viewed with pleasure again and again (short term memory is poor, of course, so maybe the patient will forget that she/he has just so recently looked at the material), providing truly therapeutic occupation. I have noted experienced caregivers taking advantage sometimes of this short term memory loss; for example, when driving must be given up (as it must, once the diagnosis of a dementia is established, or even highly suspected), confrontation with the fact that no driving will ever take place again can be avoided. The short memory span produces challenges, with the patient sometimes asking the same question again and again. This repetitious behavior is discussed below.

Indecisiveness and Problems of Judgment

Allowing an individual to attempt decision-making when it is impossible can cause frustration and anguish for both patient and caregiver, even for simple things like choosing clothing. The caregiver needs to learn the skill of "taking over" without patronizing, allowing the patient a sense of control, without distressing them by allowing them or encouraging them to go beyond their capabilities. Simple techniques, like leaving only one set of clothing in the closet, can allow the person the dignity of "choosing" his/her own clothes, without the embarrassment and argument of persuading them out of inappropriate clothing or several sets of underwear, for example. This can often be a difficult adjustment when it is the patriarch or matriarch figure of the household who is stricken. The technique is gentle firm encouragement to do a particular thing, obtaining the person's cooperation, without forcing him/her to make a decision about whether they should do the thing or not.

Disorientation/unadaptability

It is recognized that many Alzheimer's patients feel very uncomfortable in even a slightly unfamiliar environment, particularly if there are multiple other inputs (noise, crowds, etc.). This can often be assisted by companionship, assuring the person a one-on-one relationship with someone who can guide them: this is especially important for the newly hospitalized or institutionalized. Family members need to realize that what they thought of as a "treat" might be disruptive. Often the preserved insight leads the patient to refuse to go to a family event, for example, or to go on vacation, recognizing that it is too much for him. Thus necessary relocation requires careful planning and unnecessary relocation is to be avoided.

Personality Change/Disinhibition

It is a particular challenge to watch one's own relative change, particularly if some of the initial changes seem to be an exaggeration of previous personality traits and/or the early symptomatology is truly "completely unlike" the person they knew, loved and respected. Caregivers often can have problems in realizing that this is not willfulness and have a hard time emotionally disassociating themselves from the pain of seemingly unkind comments. It takes a strong, well-rested caregiver to avoid confrontation and argument, both counterproductive behaviors which will ultimately be disruptive.

Lost Daily Living Skills

Many Alzheimer's patients suffer apraxia, that is, they cannot carry out motor functions, even though they have an intact peripheral neuromuscular system. The result is progressive, the basic self-care skills are lost, and it is generally in the following order, as noted by Katz, et al. (20): bathing, dressing, toileting, transferring continence and eating. The skilled caregiver must not induce dependency by taking over too soon, and must work aggressively to recover function when it is temporarily lost, as often will happen during intercurrent illness, especially if hospitalization is involved. But to maintain true independence, simplification of some things is also very useful: Velcro fasteners, no-lace shoes, etc. Such management reduces the burden on the caregiver, and increases self-worth, independence and dignity for the patient. As any parent knows, however, it often _seems_ quicker and more effective to do something oneself, but in the long run, the caregiver who "takes over" too soon is truly creating a rod for his/her back.

Dysmobility

Wandering is problematic in some patients, but many Alzheimer's patients, particularly in the later stages of their illness, become more immobilized than they need, with a rapidly downhill progression of stiffness, dizziness on standing, reduced range of movement, and thus lost walking and transferring skills. Again, a burden is being made if efforts to keep the person mobile are not continually made. The downhill pattern can be much accelerated during intercurrent illness, especially with hospitalization. This is where the physician (who has the power to order mobilization) and the caregiver need to collaborate in ensuring that this downhill cascade does not occur. Once a late dementia patient has lost a physical skill, it will frequently not be possible to regain it. Late in the illness, the dysmobility is accelerated by the characteristic late neurological manifestations in the lower extremities-- especially increased muscular tone, sometimes amounting to frank spasticity of the lower limbs.

Aggressiveness/Catastrophic Reaction

Aggressive outbursts are very distressing and disruptive--they may produce actual injury of the caregiver, and they will make the person extremely difficult to place. Alcohol may make the outbursts worse. The classical "catastrophic reaction" is a verbal, emotional or physical out-burst, and it is generally felt to be caused by over-stimulation or too much stress, perhaps inappropriate demands for decision making or simply too much input and too many choices. Whereas the dementia patient can get into a horrendous range of difficulties just trying to rearrange objects on a shelf, or when they are all mixed up in their bed clothes, some catastrophic reactions come about because of inappropriate caregiver behaviors: sometimes in frustration, family members virtually bait the patient into this situation. The catastrophic reaction is well known to mental health nurses, and the agitation, inappropriateness and restlessness that often precede it can enable such reactions to be anticipated, and the person can be distracted and calmed. The key to the management of catastrophic reactions and aggressive outbursts is prevention, recognizing that over-stress and over-stimulation must be avoided. However, sometimes these reactions are a partial manifestation of underlying, extremely agitated, nearly psychotic thought processes in the Alzheimer's patient, and with so much agitating, distracting and meaningless mental activity going on, it is reasonable to regard this as a target symptom worthy of a major tranquilizer. More about this later.

Insomnia/"Sundowning"

Disordered sleep heads the list of factors stressing the caregiver. Elders sleep pretty badly anyway on the whole, with less deep sleep and more "awakenings," of longer duration than occur in younger patients. But in addition to an exaggeration of this in many dementia patients, there is frequently a loss of the day-night rhythm, and early recognition that this is occurring can be important in taking steps to prevent it becoming a major issue. It is very difficult to re-establish "sleep hygiene" and move the hours of sleep back to the right time when they have become established at a time several hours off every one else's pattern. The patient who regularly settles to sleep between six and seven in the evening may do so to the relief of the caregiver, but it is then inevitable that he or she will be awake by the small hours--the patient will have had enough sleep by then. Similarly, the benignity of the afternoon "nap" has to be countered by the fact that less sleep will be needed in the nighttime as a result. It is well recognized that sleep can be improved by an exercise program in the morning or early afternoon, and by calming down the evening's activities. So, frequently it is the evening that is the most stimulating and noisy time in a typical household, with nothing going on all day. Inevitably, a restless evening follows, and as the problems progress, the classic "sundowning" can sometimes progressively occur. In an established case of "sundowning syndrome" the patient becomes virtually psychotic, agitated, perhaps even hallucinated, but at least very confused, and sometimes belligerent and aggressive if crossed, not cooperating with efforts to become undressed, and certainly not able to sleep. Whether this phenomenon is related to intrinsic corticosteroid levels or whether it is a psychological manifestation of darkness falling (or whether, as is most likely, it is a combination of both) is not clear, but it is recognized that once fully established, as a regular evening pattern, whereas the efforts implied above to secure a calm evening and physically tired person are appropriate, it is frequently necessary to utilize medication. Some patients will respond to modest doses of, for example, over-the-counter diphenhydramine (Benadryl), best given in the early evening, to secure a calm evening, and occasionally the minor tranquilizers, like the benzodiazepines, can be helpful. But generally, the treatment of established "sundowning syndrome" consists of low doses of the major tranquilizers. The recommended starting doses are summarized in Table 6 (the column related to nighttime dose). As can be seen, the doses are very low compared to those used in psychiatric practice, and it may be necessary to titrate up to higher doses than this, but this is where one should start.

Demanding/Repetitive Behaviors

This can be very stressful. Some caregivers will describe "shadowing," in which the patient follows the caregiver, seeking the reassurance of the familiarity of their caregiver, and perhaps asking the same question again and again and again. The temptation is to keep answering the question, but eventually exhaustion will set in. The technique for aborting such behavior by withdrawing and coming back after a short interval with a fresh subject (which takes advantage of the reduced attention span of the patient) can be very useful.

Wandering

It is important to realize that wandering is a middle phase phenomenon, since as the patient becomes even more disorganized, they get to be unable to open even the noncomplex locks of the average front door. But wandering demands attention urgently, for the patient can get lost, there is a danger of accidents, and the difficulty and disruption of seeking a lost person and having them brought back. Walking, of course, is essential for health, and it is inappropriate to deprive Alzheimer's patients of a reasonable walking regime! At the very least, the Alzheimer's patient must wear an identity bracelet or an identity card in the shirt pocket, and often a degree of "perimeter control" such as complex locks, even in the person's own house, can be implemented (21).

Delusions/Illusions/Hallucinations/Paranoia

Misperception of things in the person's environment are often at the basis of this whole spectrum of "psychiatric"-sounding symptoms. If there is hearing impairment as well, then muttered voices (or many caregivers'--and far too many health professionals'--habit of discussing the person in front of him or in his near-hearing, on the assumption that the patient does not understand), can certainly, predictably, induce suspiciousness. But very often delusions arise from other factors, like moving shadows on the curtain. Later, spontaneous visual hallucinations can occur. In Alzheimer's, these hallucinations are generally very benign and hardly require treatment. However, the caregiver may well regard this as evidence that the person is finally "going crazy," and they may need reassurance.

ANTICIPATORY PLANNING/LEGAL AND ETHICAL ISSUES

No account of the planned management of an Alzheimer's patient would be complete without noting that a forward-looking approach must always be encouraged. One does not wish to depress the caregiver with the knowledge of the patient's certain ultimate deterioration, but issues of intensity of treatment, durable power of attorney, and guardianship and testamentary capacity, should be dealt with sooner rather than later. Decisions over DNR (do not resuscitate) status are much better taken if there is some perception of the person's own wishes incorporated into the planning.

An important part of anticipatory planning is potential institutional placement. Many family members, perhaps especially after several years of caregiving, feel terrible when their relative has to be placed in a long-term-care facility. And yet, one

TABLE 6

Some Major Tranquilizers in Dementia

Drug	Daily Dose (range)	Night Dose (range)	Parenteral (for delirium)
Thioridazine (Mellaril®) (more sedative, more anti-cholinergic, less extrapyramidal, low potency/high dose)	10-60 mg	10-50 mg (10-200 mg)	None
Haloperidol (Haldol®) (less sedative, less anticholinergic, more extrapyramidal, high potency/low dose)	2 mg (1-15 mg)	1 mg (1-10 mg)	1-2 mg
Thiothixene (Navane®) (similar side effect profile as haloperidol)	3 mg (3-45 mg)	2 mg	1-2 mg

(From Ham, R. et al. Primary Care Geriatrics: A Case-based Learning Program. John Wright. PSG. Boston, Bristol, London, 1983. (c) 1983. PSG Inc. with permission.)

so often sees, for example, younger family members or the entire rest of the family being unreasonably disrupted by the presence of an Alzheimer's patient with astonishing skilled nursing needs, which, when combined with lost awareness of the environment and even non-recognition of those close relatives around the patient, make it an unreasonable burden. Sometimes the physicians or other health professionals must be aware of their power and need to give "permission" (sometimes even more firmly, an order) to implement appropriate institutional placement. Again, anticipatory planning may be necessary, for unfortunately, the majority of areas have waiting lists for facilities, and the choices should be made rationally rather than in a crisis.

SPECIFIC MEDICATIONS FOR THE ALZHEIMER'S PATIENT

Antidepressants

Mention was made above about the concurrence of depression and dementia. In those patients in the early-to-mid phases of Alzheimer's who develop a functional deterioration, perhaps dysphoria, or perhaps just increased cognitive impairment with a sleep disorder or a diurnal variation in mood, or other suggestive but less specific symptoms such as constipation or psychomotor retardation, especially if there is a history of depressive illness, then a trial of anti-depressant may be justified. The less anticholinergic ones are generally favored. A trial starts at low doses but goes to a decent therapeutic level, maintaining that level for at least six weeks before the "trial" is terminated. It is vital that continuing observation be carried out. Once the anti-depressant starts to work, as it frequently will, it should be maintained for at least six-to-nine months, that being the normal length of an elderly depression. Many of us in this field have also seen dysphoric symptoms in the late dementia patient resolved with the use of anti-depressants, this possibly being a special place for the phenothiazine-related anti-depressant amoxapine.

Major Tranquilizers

Major tranquilizers are often regarded poorly by physician and family alike. Over-sedation must be avoided; the thought that a demented patient is being "drugged" for management purposes is anathema to most. But the fact is that, like all powerful agents, major tranquilizers can be astonishing forces for good as well as ill. The problems in their use are frequently that they are introduced at too high a dose, the dose is titrated up too fast, and, much more significantly, the target symptom is insufficiently identified to allow accurate follow-up as to the effectiveness of the medication.

The major tranquilizers really only have their place for certain target symptoms:
- aggressiveness
- sundowning syndrome
- severe mental agitation

Note that "wandering" is not on this list, nor is simple insomnia--lesser agents might help there. But the patient who is in a state of chronic agitation, distracted by near-psychotic thoughts, agitated, perhaps sulky, grumpy or frankly belligerent, tensed up like a coiled spring, ready to react, perhaps particularly if there is associated hallucinosis, does, for humanity's sake, merit careful use of a major tranquilizer. The starting doses are detailed in Table 6. The less anti-cholinergic medications, such as the high potency, more recent anti-psychotics are of course generally preferred. Careful follow-up is essential. It must be documented as to whether or not the target symptom was helped; long term treatment is hardly ever justified.

CARING FOR THE CAREGIVER

To summarize my advice to the caregiver and family:

- Realistically recognize the nature of the illness and make plans accordingly.

- Recognize your own need for help and respite: seek it, accept it, pay for it if necessary.

- Find a peer group, generally through the Alzheimer's Association: the relief of meeting others with similar problems and obtaining their advice will be invaluable.

- Communicate properly with family members: hiding how bad things are from other relatives is in the long run too stressful.

- Keep yourself well: get enough sleep, find physical outlets, keep up social contacts in your own life, and counter the tendency to withdraw yourself socially.

- Plan for your own future: the Alzheimer's patient will likely die first, and you will have life afterwards; plan for it socially and financially.

- Become informed about Alzheimer's, so that you can anticipate problems and have strategies planned.

- Plan the financial and legal aspects, including the will. Seek professional advice if necessary, and get a power of attorney established.

- Most positive, and most important: continually work to find and exercise the preserved function of the patient. This will reduce the burden, increase the quality of your existing relationship with the patient, and make your relative's life more rewarding.

SUMMARY

I have tried to set a keynote--an emphasis on taking a positive approach to the multiple problems that Alzheimer's presents to the patient, the caregiving family, the health professionals, the health care system and society-at-large. The attempt is to preserve as much function as possible, for as long as possible, and thus to maintain the person in his/her own home, in a familiar environment, the environment in which he/she will be least confused and most productive and calm, for as long as possible; in the environment which is most economical, so that as much as possible of the person's capability and person can still be enjoyed by the family members, despite the presence of this illness. One does not wish to make light of this illness and the tragedy that it represents for many of those whom it afflicts, and particularly for the family members, who are not blessed with the gift of the frequently diminished insight the patient has as the illness progresses. However, pleasure, joy, love, affection, and an appreciation of excitement, reminiscence and beauty, and particularly love and affection and many other life-enhancing qualities are still there to be enjoyed by patient and family member alike. To achieve this, the family member must be well informed, skilled, and, above all, well rested and energetic, if he or she is to be the means of reducing the impact of Alzheimer's on the patient, the family and our society.

REFERENCES

1. Alzheimer, A. (1907). Uber Eine Eigenartige Erkrankung Der Hirnrinde. *All Z Psychiatry, 64:*146-148.
2. Siegal, J. S. (1986). Demographic perspectives on the long lived society. *Daedalus. 115,* 72-117.
3. Gallagher, J. C., Melton, L. J., Riggs, B. L., Bergstrath, E. (1980). Epidemiology of fractures of the proximal femur in Rochester, Minnesota. *Clin. Orth., 1150:*163-171.
4. Breitner, J. C. S. (1987). The new genetics of Alzheimer's disease. Clinical Report on Aging. *AGS., 1*(3) 1-5.
5. Ham, R. J. (1987). Alzheimer's and the family. In *Geriatric Medicine Annual 1987:* Ham, R. J. (ed). Medical Economics Books, Oradell, N. J.
6. Williamson, J. (1965). Old people at home: Their unreported needs. *Lancet,* 1117-1120.
7. Folstein, M. F. (1975). Mini-mental state: A practical method for grading the cognitive state of patients for the clinician. *J. Psychiatr. Res., 12*(3):189-198.
8. Katzman, R. (1986). Alzheimer's disease. *N. Engl. J. Med., 314*(15):964-973.
9. Ham, R. J. (1988). Functional assessment of the elderly patient. In *Clinical Aspects of Aging*, Third Edition. Reichel, W. (ed), Williams and Wilkens, Baltimore, MD.
10. *DSM-III R: (Diagnostic and Statistical Manual of Mental Disorders)* (1987), Third Edition, Revised. American Psychiatric Association, Washington, D.C.
11. Omnibus Budget Reconciliation Act, 1987. U.S. Government Printing Office, Washington, D.C., 1989.
12. Ham, R. J., Smith, M. R. (1983). The confused patient. In *Primary Care Geriatrics: A Case-Based Learning Program*. Ham, R. J. (ed). John Wright-PSG. Boston, Mass. (distrib. Yearbook Medical, Chicago, IL).
13. Larson, E. G., Reifler, B. V. (1984). Sumi S. M. (1983). Dementia in elderly out-patients: A prospective study. *Ann. Int. Med., 100:*417-423.
14. Clarfield, M. (1988). The reversible dementias: Do they reverse? *Ann. Int. Med., 109:*476-486.

15. Barry, P. P., Moskowitz, M. A. (1988). The diagnosis of reversible dementia in the elderly: A critical review. *Arch. Intern. Med.*, *148*:1914-1918.
16. Solomon, D. H. (1988). Geriatric assessment: Methods for clinical decision making. *JAMA.*, *259*(16):2450-2452.
17. McKhann, G., Drackman, D., Folstein, M., (1984). Clinical diagnosis of Alzheimer's disease. *Neurology, 34*:939.
18. Wells, C. E. (1979). Pseudodementia. *Am.J. Psychiatry, 136*:7.
19. Coffey, C. E. (May, 1989). Structural brain changes on MRI in the elderly: Relationship to late-onset depression. Presented at American Geriatrics Society, 46th Annual Meeting, Boston, MA.
20. Katz, S. (1970). Progress in development of the index of ADL. *Gerontologist, 10*(1):20-30.21. Winograd, C. H., Harvik, L. F. (1986). Physician management of the demented patient. *JAGS, 34*(4):295-308.
21. Winograd, C. H., Harvik, L. F. (1986). Physician management of the demented patient. *JAGS, 34(4)*:295-308.

Illustration 1

Mini-Mental State Examination
(Folstein, 1975)

I. **Orientation (Maximum score: 10)**
Ask "What is today's date?" Then ask
specifically for parts omitted, such
as "Can you also tell me what season
it is?"

Date (e.g., January 21)................ 1 ____
Year.. 2 ____
Month.. 3 ____
Day (e.g., Monday)....................... 4 ____
Season .... 5 ____

Ask "Can you tell me the name of this
hospital?"
"What floor are we on?"
"What town (or city) are we in?"
"What county are we in?"
"What state are we in?"

Hospital .... 6 ____
Floor... 7 ____
Town/City................................. 8 ____
County...................................... 9 ____
State... 10 ____

II. **Registration (Maximum score: 3)**
Ask the patient if you may test his
memory. Then say "ball," flag,"
"tree" clearly and slowly, allowing
about one second after each. After you
have said all three words, ask the
patient to repeat them. This first
repetition determines the score (0-3),
but continue to say them (up to six
trials) until the patient can repeat
all three words. If he does not
eventually learn all three, recall cannot
be meaningfully tested.

"ball".. 11 ____
"flag".. 12 ____
"tree".. 13 ____

Number of trials:_____

III. **Attention and calculation (maximum
score: 5)**
Ask the patient to begin at 100 and count
backwards by 7. Stop after five
subtractions (93,86,79,72,65). Score
one point for each correct number.

"93".. 14 ____
"86".. 15 ____
"79".. 16 ____
"72".. 17 ____
"65".. 18 ____

or

If the subject cannot or will not perform
this task, ask him to spell the word
"world" backward (D,L,R,O,W). Score one
point for each correctly placed letter,
e.g., DLORW = 3. Record how the patient
spelled "world" backward: _____
                    D L R O W

Number of correctly placed
letters............................... 19 ____

IV. **Recall (Maximum score: 3)**
Ask the patient to recall the three words
you previously asked him to remember
(learned in Registration)

"ball".. 20 ____
"flag".. 21 ____
"tree".. 22 ____

V. **Language (Maximum score: 9)**
Naming: Show the patient a wristwatch
and ask "What is this?" Repeat for a
pencil. Score one point for each item
named correctly.

Watch .... 23 ____

Pencil....................................... 24 ____

Repetition: Ask the patient to repeat
"No if's, and's or but's." Score one
point for correct repetition.

Repetition................................. 25 ____

Three-stage command: Give the patient
a piece of blank paper and say "Take the
paper in your right hand, fold it in half
and put it on the floor." Score one
point for each action performed correctly.

Takes in right hand .... 26 ____

Folds in half .... 27 ____
Puts on floor .... 28 ____

19

Illustration 1 (continued)

Reading: On a blank piece of paper, print the sentence "Close your eyes" in letters large enough for the patient to see clearly. Ask the patient to read it and do what it says. Score correct only if he actually closes his eyes.

Closes eyes.....................................29_____

Writing: Give the patient a blank piece of paper and ask him to write a sentence. It is to be written spontaneously. It must contain a subject and verb and make sense. Correct grammar and punctuation are not necessary.

Write sentence ....30_____

Copying: On a clean piece of paper, draw intersecting pentagons as illustrated, each side measuring about 1 inch, and ask the patient to copy it exactly as it is. All 10 angles must be present and two must intersect to score 1 point. Tremor and rotation are ignored.

Draws pentagons.........................31_____

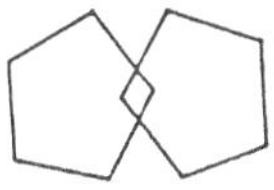

Score: Add number of correct responses. In Section III, include items 14 through 18 or item 19, not both (Maximum total score: 30).

Total Score_____

Level of consciousness: _____coma _____stupor
_____drowsy _____alert

(Reprinted with permission from Folstein, M. F., Folstein, S. E., McHugh, P.R. Mini-mental state: A practical method for grading the cognitive state of patients for the clinician. J Psychiatry Res 1975; 12(3):189-98. (c) Pergamon Press, Ltd.)

IN SEARCH OF THE ETIOLOGY OF ALZHEIMER'S DISEASE

John A. Edwards, M.D., F.A.C.P., F.R.C.P.

Professor, Departments of Medicine and Family Medicine
State University of New York at Buffalo
Buffalo, New York

Alzheimer's disease is defined by a characteristic neuropathology consisting of neurofibrillary tangles made up of intraneuronal paired helical filaments and straight filaments, and senile plaques made up of a core of extracellular amyloid fibrils surrounded by dystrophic neurites and glial cells, the accumulation of the same amyloid fibrils in cerebral and meningeal microvessels and loss of neurones. This neuropathology predominantly affects the frontal, parietal and temporal cortex, and hippocampus.

The etiology of Alzheimer's disease is not known. Hence, it is not surprising that there is no preventive or curative treatment. This situation is likely to persist until the cause or causes of Alzheimer's disease are elucidated.

OBSTACLES TO THE ELUCIDATION OF ETIOLOGY

Making a definite diagnosis

Unless an autopsy or brain biopsy is done, the clinical diagnosis of Alzheimer's disease remains a probable one. Even after extensive clinical, laboratory and radiological investigation the diagnosis of Alzheimer's disease is not one hundred percent accurate. The difficulty in making the diagnosis of Alzheimer's disease is a particular obstacle in epidemiologic and genetic studies. The epidemiologic search for possible etiological factors using case control studies is handicapped by the lack of diagnostic certainty and the absence of a simple, inexpensive, specific test for the disease. Comparisons between the frequency of Alzheimer's disease in different cultural, ethnic and geographically defined groups customarily require the study of very large numbers of individuals. Such studies are made difficult, expensive and very time consuming by the extensive investigations required to make a diagnosis of probable Alzheimer's disease. In family studies it is clearly difficult to be sure whether a relative who died many years ago had Alzheimer's disease.

Inaccessibility of the brain

The more accessible an organ or tissue is, the easier it is to discern the etiology and pathogenesis of diseases affecting that organ or tissue. It is no accident that rapid advances in our knowledge of hematological disorders have been made and continue to be made. It is due in part to the ease with which blood can be obtained for study. The brain is rendered relatively inaccessible by the bony box it is contained in and by the blood brain barrier. This inaccessibility, although of

New Directions in Understanding Dementia and Alzheimer's Disease,
Edited by T. Zandi and R. J. Ham, Plenum Press, New York, 1990

21

evolutionary advantage in helping protect this vital organ from damage, represents an obstacle to the study of brain disease. Obviously biopsy of organs such as the liver and kidney is a much easier proposition than biopsy of the brain. To some extent, recent technological advancements such as computerized axial tomography, magnetic resonance imaging and positron emission tomography are helping to overcome the inaccessibility of the brain.

Lack of a suitable animal model

Although there are some reservations in the extrapolation of data obtained in animals to man, these are usually more than offset by the ease of experimental manipulation. Clearly there are studies that could be done in laboratory animals that could not be done in human subjects. Hence, it would be of great benefit to have an animal model of Alzheimer's disease. Unfortunately, there is no good animal model of the disease. Although selective areas of the brain of laboratory animals can be damaged or destroyed by physicochemical means, this type of model does not simulate a chronic neurodegenerative disease such as Alzheimer's that affects several neurotransmitter systems.

Distinguishing between basic lesion and epiphenomenon

A large number of extra neuronal abnormalities have been found in patients with Alzheimer's disease involving red blood cells, lymphocytes, granulocytes, platelets and fibroblasts (1). Red blood cell abnormalities include increased choline content, decreased cholinesterase and increased sodium-lithium counter transport. Lymphocyte findings have included enhanced suppressor cell activity, decreased natural killer cell activity, increased radio-sensitivity and increased chromosome aberrations. Granulocytes have been shown to have decreased mobility and platelets to have decreased phosphofructokinase activity. Abnormalities of fibroblasts have included microtubular defects, deficient DNA strand break repair, and defects in calcium metabolism and mitochondrial function.

All the abnormalities outlined above have been associated with an overlap in values between patients with Alzheimer's disease and control subjects. Hence, none of them are useful from the diagnostic point of view.

A critical question about the above findings is whether any one of them represents the basic etiological defect or whether all of them are the secondary consequence of some other yet to be discovered basic abnormality. Clearly, not all of the above abnormalities can have a basic etiological significance. It is possible that some or all of them are epiphenomena resulting from the effect of Alzheimer's disease on the neuro endocrine system or nutritional status.

Probable etiological heterogeneity

Just because Alzheimer's disease is defined by one single neuropathological picture does not necessarily mean that it is one single disease. There is increasing evidence of clinical and biological heterogeneity (2). There is much intersubject variability in behavioral features, cognitive dysfunction, associated motor deficits, age of onset, and rate of progression of the disease. Some patients have evidence of predominant left hemisphere involvement whilst other patients have evidence of predominant right hemisphere involvement. Some patients are agitated, others are not. Some patients have myoclonus or extra pyramidal defects, others do not. Some patients have an extra chromosome 21 (Down's syndrome), most do not.

Perhaps the most persuasive evidence of etiological heterogeneity is provided by the early onset familial form of Alzheimer's disease and the Down's syndrome associated form of the disease. Although there remains some controversy on whether the early onset, familial form of the disease and the late onset form of the disease are truly distinct and separate entities, these two forms of the disease tend to "breed true" in the families studied; i.e., the affected family members of patients with early onset disease tend to have early onset disease and the affected family members of patients with late onset disease tend to have late onset disease.

A point of interest with respect to the Down's syndrome associated form of Alzheimer's disease is that not all patients over the age of forty years have clinical evidence of dementia although they all have the characteristic neuropathological features of the disease. Furthermore, this cannot be attributed to the difficulties in recognizing dementia in a mentally retarded individual. This observation raises the issue of the exact relationship between the neuropathology of Alzheimer's disease and its clinical manifestations.

Distinguishing cause and effect

Of all the obstacles to the elucidation of the etiology of Alzheimer's disease, that of distinguishing between cause and effect is probably the greatest. It is always difficult to determine whether an abnormality is the cause or the result of the disease. This difficulty tends to be magnified when studying patients at or towards the end of a long disease process.

Since cause precedes effect, an Alzheimer's disease associated abnormality would be more likely to be causal if it were found in persons with the disease who had not yet become symptomatic. Hence, two prime groups for studies on the etiology of the disease would be the offspring and sibs of patients with autopsy proven, presenile, familial Alzheimer's disease and individuals with Down's syndrome below the age of forty years. With regard to the former group, it would be expected that fifty percent of them would be carrying the gene for Alzheimer's disease. Hence any abnormality found in this group of subjects should have a bimodal frequency distribution if it were of etiological significance.

ETIOLOGICAL HYPOTHESIS

Environmental toxins

Although no environmental toxin has been identified as a definite cause of Alzheimer's disease, there are two interesting precedents for toxin induced neurodegenerative disease.

One of these involves amyotrophic lateral sclerosis-Parkinsonism-dementia among the Chamorro population of the Western Pacific islands of Guam and Rota (3). The decline in the high incidence of this disorder, taken together with the lack of evidence for a viral or hereditary cause, suggested that an environmental factor associated with the Chamorro culture might be gradually disappearing. One possible such factor was the seed of the Cycas Circinalis plant, a traditional source of food and medicine that had been used less since the Americanization of Guam following the second World War. The seed of Cycas Circinalis contains an amino acid (N-methyl amino-L-alanine), a low potency convulsant. When this amino acid was fed to Macaques, they developed corticomotor neuronal dysfunction, Parkinsonian features and behavioral abnormalities along with degenerative changes in motor neurones in the cerebral cortex and spinal cord. The experimental and epidemiological evidence taken together strongly suggest that Cycad exposure plays an important role in the etiology of the Guam disease.

A second precedent involves the pyridine derivative MPTP that selectively destroys cells in the substantia nigra and hence causes Parkinsonism (4). This connection between MPTP, which first surfaced in 1977, has led to the development of a very useful animal model of the disease. It is not MPTP itself that is toxic, but MPPT, a product of the action of monoamine oxidase B on MPTP. The glial cells surrounding the substantia nigra neurones convert MPTP to MPPT (1-methyl-4-phenyl pyridine) which is then selectively accumulated by the catecholamine pump of these neurones. The MPPT binds to neuromelanin and is then slowly released and damages the dopamine neurones. Although MPTP is not a naturally occurring chemical, it is closely related to a large number of compounds which do occur naturally. Could it be that a naturally occurring compound that is toxic to cholinergic neurones is a cause of Alzheimer's disease? Currently there is no evidence to support this speculation.

The possible role of aluminum in the etiology of Alzheimer's disease remains an unsettled issue (5). It is now well established that aluminum accumulates in neuro-fibrillary tangles. It seems unlikely that aluminum has an etiological role because:

    a)   It is one of the most ubiquitous elements in our environment, hence we are all exposed to it.

    b)   There are no reported epidemiological studies linking exposure to aluminum and Alzheimer's disease.

    c)   The accumulation of aluminum in neurofibrillary tangles is not specific for Alzheimer's disease.

    d)   The neuronal lesion produced by the injection of aluminum salts into the brain of experimental animals is not exactly the same as the neuronal lesions of Alzheimer's disease.

Infectious agents

As with environmental toxins, there are precedents for infectious agents causing neurodegenerative disease in both animals and man. Creutzfeld Jacob disease and Kuru are two human neurodegenerative diseases caused by slow viruses or prions. Both diseases can be transmitted to experimental animals using cell-free filtrates (6).

Prions are highly unconventional agents since they are highly resistant to ultraviolet light and they lack detectable nucleic acid. With regard to Alzheimer's disease, it is of interest to note that purified preparations of prions have a rod-like structure and properties similar to the amyloid fibrils found in the core of senile plaques and the walls of cerebral blood vessels (6). Despite this observation, at present there is no convincing evidence that Alzheimer's disease can be transmitted to experimental animals using brain homogenate (7).

However, recent work has shown that when buffy coat leukocytes from one patient with Alzheimer's disease and four relatives of patients with Alzheimer's disease were injected into the brain of hamsters, lesions of a spongiform encephalopathy, similar to those of experimental Creutzfeld Jacob disease developed 196 to 517 days following inoculation. Furthermore, when leukocytes from the affected hamsters were in turn injected into the brain of other hamsters, a spongiform encephalopathy was again produced. These transmission results in a small number of individuals raise the possibility that CJD-like agents may be involved in at least some forms of Alzheimer's disease (8).

Genetic factors

In the final analysis, all human disease is due to genetic factors, environmental factors and their interaction, and Alzheimer's disease is no exception. However, the relative importance or strength of genetic factors varies from disease to disease. There is little doubt that the early onset, familial form of Alzheimer's disease is an autosomal dominant trait. Furthermore, it has been shown by recent linkage studies utilizing various restriction fragment length polymorphic markers that the gene determining this form of the disease occupies a locus on the long arm of chromosome 21 (9), and that the gene is not allelic with the one determining the synthesis of the precursor of B-amyloid protein (A4 protein) (10,11).

The controversy as to whether the early onset and late onset forms of Alzheimer's disease are the same or different diseases has already been mentioned in an earlier section of this chapter. One of the main problems in trying to resolve this issue is the limitations of clinical genetic studies in a late onset disorder and the difficulty in making the diagnosis of Alzheimer's disease in persons long dead. For example, if a relative of a patient with Alzheimer's disease dies at the age of sixty-five of a myocardial infarction, there is no way to determine whether that relative would have developed Alzheimer's disease if he or she had lived longer. Despite the above problems and difficulties, it is beginning to look as though the early onset and late onset forms of the disease are separate and distinct entities. The evidence supporting this tentative conclusion is:

a) The early and late onset forms of Alzheimer's disease tend to "breed true" in families.
b) The early onset form of the disease generally runs a more rapid downhill course than the late onset form of the disease.
c) Studies of identical twins with late onset Alzheimer's disease show a lack of concordance even when observations are continued over several years.
d) Linkage studies of the early onset form of the disease have shown that the gene determining the disorder is located on chromosome 21; no such linkage studies have been performed on subjects with the late onset form of the disease.

Metabolic defect

Some of the metabolic defects found in non-neuronal tissue from patients with Altzheimer's disease have already been mentioned in an earlier section of this paper. Aging and Alzheimer's disease lead to alterations in several biochemical processes of cultured skin fibroblasts (12). Total bound calcium increases in normal aging (+52%) and more so in Alzheimer's disease (+197%). Processes that require mitolchondrial function, such as glucose and glutamine oxidation, decline in cells from aged donors (-25%) and more so in Alzheimer's disease (-46%). Furthermore, biosynthetic processes that depend on mitochondrial function such as glucose and glutamine incorporation into protein and lipids, parallel the oxidative decreases. These metabolic defects in non-neuronal tissue from Alzheimer's disease patients raise some important questions:

a) Are the deficiencies in mitochondrial function or calcium homeostatis etiological or secondary to alterations in the cells' neuro-endocrine environment induced by Alzheimer's disease? Although tissue culture allows factors such as nutritional, clinical, and neuro-endocrinological status to be minimized, it does not completely eliminate the possibility that such factors might be responsible for the defects found.

b) If the deficiency in mitochondrial function or calcium homeostasis represented the primary basic lesion of Alzheimer's disease, why are the manifestations of the lesion confined to particular brain neurons? Perhaps the manifestations are not confined to the brain. Perhaps the weight loss that occurs in Alzheimer's disease is a manifestation of mitochondrial dysfunction. However, if the consequences of mitochondrial dysfunction are confined to certain brain neurons, then this would imply that these neurons have an innate hypersusceptibility to alterations in calcium homeostasis or mitochondrial dysfunction and a limited ability for regeneration.

Recent studies have shown that fibroblast cell lines from Alzheimer's subjects, when grown under conditions favoring the expression of neuronal properties, stained immunocytochemically for paired helical filaments, whereas none of the controls did so. However, when cell lines from control subjects were grown in the presence of the mitochondrial uncoupler CCCP, then they did stain immunocytochemically for paired helical filaments (13). These findings suggest that neurofibrillary tangles might develop as a result of mitochondrial dysfunction. Whether mitochondrial dysfunction is the cause of Alzheimer's disease or a consequence of the disease remains to be seen.

<u>Defective DNA repair</u>

The finding of deficient DNA repair in strains of fibroblasts and of lymphoblasts from patients with several types of primary neuronal degeneration has led to the hypothesis that efficient DNA repair is required to maintain the functional integrity of the human nervous system (14). Fibroblasts and lymphoblasts from patients with Alzheimer's disease have reduced survival after X-irradiation and after treatment with the radiomimetic alkylating agent N-methyl-N-nitro-N-nitrosoguanidine (MNNG) (15). DNA strand breaks, resulting from treatment with the alkylating agents MNNG and methyl-methanesulfonate (MMS) are repaired more slowly by Alzheimer's disease fibroblasts than in cells from age-matched controls (16,17). Studies of the level of O^6 methylguanine DNA methyltransferase, a key enzyme in the repair of DNA alkylation damage, have failed to show any significant difference between lymphocytes from patients with Alzheimer's disease and cells from age-matched controls (18). The significance of the demonstrated defects in DNA repair in non-neuronal tissue is difficult to assess for reasons already discussed in earlier sections of this chapter.

AMYLOID RELATED RESEARCH

The senile plaque with its core of amyloid fibrils is the most specific of the neuropathological abnormalities found in Alzheimer's disease. Senile plaques are found only in this disease, dementia pugilistica, and in small numbers in the aging brain of humans and other mammals. This specificity has been in part the reason for the research focus on amyloid over the last few years.

This research received a major stimulus in 1987 with the cloning of a complementary DNA (cDNA) probe for the gene encoding the synthesis of beta amyloid protein (A4 protein) (19,20). Once the A4 protein had been purified and partially sequenced (21,22), it was possible to synthesize oligonucleotide probes that were used to examine a cDNA library derived from human brain. Southern blot analysis of the positive clones so identified led to the production of the cDNA probe described above. The examination of mouse/man somatic cell hybrids with this cDNA probe led to the discovery that the gene encoding the A4 protein was located on chromosome 21. Further investigation using the cDNA probe also led to the discovery that the gene encoding the A4 protein was present and transcribed in multiple human tissues including brain, kidney, heart, spleen, liver and lymphocytes as well as in tissue from several non-human species including mouse, rat, cow, hamster, lobster and drosophila. Other studies using the cDNA probe showed that the A4 protein was a fragment (approximately 40 amino acids) of a much larger precursor protein (pre A4).

These initial findings raised a number of intriguing questions. One was whether the gene on chromosome 21 encoding the A4 protein was allelic with the gene on the same chromosome that was responsible for the early onset, familial form of Alzheimer's disease. As a result of classical genetic analysis, it has now been shown that the two above genes are not allelic and that they are no closer to each other than approximately eight million bases (10,11).

Another question concerned the structure and function of the pre A4 protein. Since it had been shown that the gene encoding the pre A4 protein had been highly conserved during the course of evolution, it followed that the pre A4 protein must have some important function. It now appears that the pre A4 protein is inserted in the cell membrane with the A4 portion (beta amyloid protein) partially buried in the membrane (23). Although the function of the pre A4 protein is not known, there are two leading possibilities. One is that it might help to establish or maintain connections between nerve cells at the neuronal terminal. This is consistent with the finding that the pre A4 protein is made by neurones and rapidly transported down axons (24). Another possible function of the pre A4 protein is the maintenance of neuronal growth or survival. This hypothesis is consistent with the finding that the first 28 amino acids of the A4 protein enhance the survival of brain neurons maintained in culture (25).

Another important question is what are the sequence of events between the synthesis of the pre A4 protein and the final deposition of amyloid fibrils in the core of senile plaques and cerebral blood vessels. At present it is not clear whether the pre A4 protein is synthesized predominately outside the brain and transported to the brain across the blood brain barrier, or whether it is synthesized predominantly in the brain itself. Results of studies of the level of pre A4 protein by radioimmunoassay indicate that it is 30-fold greater in serum than cerebrospinal fluid (26), and that it is 1.5 -fold greater in the serum of patients with Down's syndrome than both normal controls and patients with non-Down's syndrome associated Alzheimer's disease (27). Furthermore, the deposition of the A4 protein occurs in meningeal vessels outside the brain, a site that is similar to that of systemic amyloidosis known to originate from circulating precursors. The above findings suggest that some A4 protein deposits could originate from a circulating source. On the other hand, the restriction of amyloid fibril deposition to the brain and the abundant formation of pre A4 protein messenger RNA's by neurons suggests a neuronal origin for the amyloid fibril deposition in senile plaques (28).

Irrespective of whether the pre A4 protein is produced predominantly outside the brain or within the brain, the A4 protein is first cut out of the precursor pre A4 protein by proteases. The possible role of protease inhibitors in this process are not presently clear (24). The protease inhibitor alpha 1-antichymotrypsin is consistently present in senile plaques (29). Whether this anti-enzyme contributes to plaque formation in Alzheimer's disease by inhibiting the pre A4 degrading proteases is not known. The pre A4 protein itself can be produced both with or without and additional amino acid sequence that acts as a protease inhibitor (24). Whether this Kunitz type of inhibitor plays any part in stimulating or retarding plaque formation is again not known. Diffuse deposition of the A4 protein commonly occurs in the molecular cortex of the cerebellum, a part of the brain that is not usually involved in Alzheimer's disease. However, this diffuse deposition does not lead to any neuritic or glial response (30). Hence, local factors must be operative in the frontal, parietal and temporal cortex that lead to the aggregation of A4 protein molecules into amyloid fibrils.

In Down's syndrome, it appears that the overproduction of pre A4 protein, due to a gene dosage effect, is the main factor leading to the formation of senile plaques. Studies to date fail to provide any evidence for a similar phenomenon due to gene duplication occurring in non-Down's syndrome associated Alzheimer's disease (31,32). There is also no evidence to suggest that a mutated form of pre A4 protein is produced in late onset Alzheimer's disease. Currently most attention is being focused on the possibility that either an abnormality in protease function, a defect in protease inhibition or an abnormality involving local factors in the cerebral cortex is responsible for the deposition of amyloid fibrils in the senile plaques of Alzheimer's disease.

One last question is whether the senile plaques and their contained amyloid fibrils are the cause or the consequence of Alzheimer's disease. Earlier in this paper it was mentioned that cause precedes effect. In this regard it is important to note that young adults with Down's syndrome have diffuse deposits of A4 protein as early as the third decade and that these deposits precede the development of senile plaques and neurofibrillary tangles (27,33,34). Since all patients with Down's syndrome who survive past the age of forty years have the characteristic neuropathology of Alzheimer's disease, it can be concluded that the amorphous deposits of A4 protein probably represent the earliest abnormality in the brain of patients with Alzheimer's disease or Down's syndrome (28). This is persuasive evidence in favor of the supposition that an abnormality involving the A4 protein is causally related to the development of Alzheimer's disease. Further research on the relationship of amyloid to Alzheimer's disease holds the promise of coming to an understanding of the etiology of this tragic disease and the devisement of preventive and curative treatment.

REFERENCES

1. Blass, J. P. and Zemcov, A. (1984). Alzheimer's disease: A metabolic systems degeneration? *Neurochem.Path.,* 2:103-114.
2. Friedland, R. P. (moderator), Koss, E., Haxby, J. V., Grady, C. L., Luxenberg, J., Shapiro, M. B., and Kay, J. (discussants) (1988). Alzheimer's disease: Clinical and biological heterogeneity. *Ann. Intern. Med.,* 109:298-311.
3. Spencer, P. S., Nunn, P. B., Hugon, J., et al. (1987). Guam amyotrophic lateral sclerosis- Parkinsonism-dementia linked to a plant excitant neurotoxin. *Science, 237:*517-522.
4. Lieberman, A. N. (1987). Update on Parkinson's disease. *New York State J. Med.,* 87:147-152.
5. Perl, D. P. and Pendlebury, W. W. (1986). Aluminum neurotoxicity-potential role in the pathogenesis of neurofibrillary tangle formation. *Can. J. Neurol. Sci.,* 13:441:445.
6. Prusiner, S. B. (1987). Prions and neurodegenerative diseases. *New Engl. J. Med., 317:*1571-1581.
7. Corsellis, J. A. N. (1986). The transmissibility of dementia. *Brit. Med. Bull.,* 42:111-114.
8. Manuelidis, E. E., DeFigueiredo, J. M., Kim, J. H., et al. (1988). Transmission studies from blood of Alzheimer disease patients and healthy relatives. *Proc. Natl. Acad. Sci. USA, 85:*4898-4901.
9. St. George-Hyslop, P. H., Tanzi, R. E., Polinsky, R. J., et al. (1987). The genetic defect causing familial Alzheimer's disease maps on chromosome 21. *Science, 235:*885-889.
10. Van Broeckhoven, C., Genthe, A. M., Vandenberghe, A., et al. (1987). Failure of familial Alzheimer's disease to segregate with the A4 amyloid gene in several European families. *Nature* (London), *329:*153-155.
11. Tanzi, R. E., St. George-Hyslop, P. H., Haines, J. L., et al. (1987). The genetic defect in familial Alzheimer's disease is not tightly linked to the amyloid B protein gene. *Nature* (London), *329:*156-157.
12. Peterson, C. and Goldman, J. E. (1986). Alterations in calcium content and biochemical processes in cultured skin fibroblasts from aged and Alzheimer donors. *Proc. Natl. Acad. Sci. USA, 83:*2758-2762.
13. Baker, A. C., Ko, L. W., Shea, R. K. F., et al. (1988). Studies of "neuronal" and "Alzheimer" antigens in skin cells. *Alzheimer's Disease and Associated Disorders,* 2:178.
14. Robbins, J. II. (1983). Hypersensitivity to DNA-damaging agents in primary degeneration of excitable tissue. In Bridges, B. A. and Friedberg, E. C. (eds), *Cellular Responses to DNA Damage.* Alan R. Liss, New York.
15. Robbins, J. H., Otsuka, F., Tarone, R. E., et al. (1983). Radiosensitivity in Alzheimer's disease and Parkinson's disease. *Lancet, 1:*468-469.
16. Li, J. C. and Kaminskas, E. (1985). Deficient repair of DNA lesions in Alzheimer's disease fibroblasts. *Biochem. Biophys. Res. Comm., 129:*733-738.
17. Robinson, S. H. and Bradley, E. G. (1985). Impaired DNA repair replication in Alzheimer's disease cells. In Hutton, J. T. and Kenney, A. D. (eds), *Senile Dementia of the Alzheimer's Type.* Alan R. Liss, New York.
18. Edwards, J. A., Wang, L. G., Setlow, R. B., and Kaminskas, E. (1989). Lymphocyte O^6-methylguanine DNA methyltransferase in the elderly with and without Alzheimer's disease. *Mut. Res., 219:*267-272.

19. Goldgaber, D., Lerman, M. I., McBride, O. W., et al. (1987). Characterization and chromosomal location of a cDNA encoding brain amyloid of Alzheimer's disease. *Science, 235*:877-880.

20. Tanzi, R. E., Gusella, J. F., Watkins, P. C., et al. (1987). Amyloid B protein gene: cDNA, mRNA distribution, and genetic linkage near the Alzheimer locus. *Science, 235*:880-884.

21. Glenner, G. G. and Wong, C. E. (1984). Alzheimer's disease: Initial report of the purification and characterization of a novel cerebrovascular amyloid protein. *Biochem. Biophys. Res. Comm., 120*:885-890.

22. Masters, C. L., Simms, G., Weinman, N. A., et al. (1985). Amyloid plaque core protein in Alzheimer's disease and Down's syndrome. *Proc. Natl. Acad. Sci. USA, 82*:4245-4249.

23. Kang, J., Lemaine, H. G., Unterbeck, A., et al. (1987). The precursor of Alzheimer's disease amyloid A4 protein resembles a cell surface receptor. *Nature* (London), *325:*733-736.

24. Marx, J. L. (1989). Brain protein yields clues to Alzheimer's disease. *Science, 243*:1664-1666.

25. Whitson, J. S., Selkoe, D. J., Cotman, C. W. (1989). Amyloid B protein enhances the survival of hippocampal neurones in vitro. *Science, 243*:1488-1490.

26. Pardridge, W. M., Microvessels from Alzheimer's disease brain: Biochemistry of amyloid peptides, pp. 48-51. In Vinters, H. V., moderator (1988). Brain amyloid and Alzheimer's disease. *Ann. Intern. Med., 109*:41-54.

27. Rumble, B, Retallack, R., Hilbich, C., et al. (1989). Amyloid A4 protein and its precursor in Down's syndrome and Alzheimer's disease. *New Engl. J. Med., 320*:1446-1452.

28. Selkoe, D. J. (1989). Aging, amyloid, and Alzheimer's disease. *New Engl. J. Med., 320*:1484-1487.

29. Abraham, C. R., Selkoe, D. J., and Potter, H. (1988). Immunochemical identification of the serine protease inhibitor alpha 1-antichymotrypsin in the brain amyloid deposits of Alzheimer's disease. *Cell, 52*:487-501.

30. Joachim, C. L., Morris, J. and Selkoe, D. J. (1989). Diffuse senile plaques occur commonly in the cerebellum in Alzheimer's disease. *Am. J. Pathol., 135*:309-319.

31. St. George-Hyslop, P. H., Tanzi, R. E., Polinsky, R. J., et al. (1987). Absence of duplication of chromosome 21 genes in familial and sporadic Alzheimer's disease. *Science, 238*:664-666.

32. Tanzi, R. E., Bird, E. D., Latt, S. A., et al. (1987). The amyloid B gene is not duplicated in brains from patients with Alzheimer's disease. *Science, 238*:666-669.

33. Yamaguchi, H., Hirai, S., Morimatsu, M., et al. (1988). A variety of cerebral amyloid deposits in the brain of the Alzheimer's-type dementia demonstrated by beta protein immunostaining. *Acta Neurolpathol., 76:*541-590.

34. Giacone, G., Tagliavini, F., Linoli, G., et al. (1989). Down patients: Extracellular preamyloid deposits precede neuritic degeneration and senile plaques. *Neuro. Sci. Lett., 97*:323-328.

ALZHEIMER'S DISEASE: THEORIES OF CAUSATION

Walter G. Bradley, D.M., F.R.C.P

Chair, Department of Neurology
College of Medicine
The University of Vermont
Burlington, VT

INTRODUCTION

Alzheimer's disease will be the epidemic of the twenty-first century (1,2). The generally accepted figures are that 5% of individuals of the age of 65 have severe dementia, and another 10% have moderate dementia, with 30% having developed dementia if they live to the age of 90. The recent Boston study suggested that the figures were even higher, with 3% of those age 65 to 74 having probable Alzheimer's disease, 18.7% of those 75 to 84, and 47.2% of those over 85 having probable Alzheimer's disease (3). About 75% of patients with a clinical diagnosis of senile dementia prove to have Alzheimer's disease, while a number of other disorders, such as multi-infarct dementia and other neurological degenerations, underlie the remaining proportion. Alzheimer's disease is therefore going to be the major health care burden in the coming decades, both in terms of personal and family stress and of national health care costs. It is unnecessary, therefore, to emphasize the urgent need for an understanding of the cause of the disease and its treatment.

There are many theories of the causation of Alzheimer's disease, all with some degree of support from experimental evidence. This paper will review the current status of several of these theories.

NEURONAL LOSS

Alzheimer originally indicated that there was a loss of cortical neurons in this disease (4,5), though quantitative demonstration of this has only recently been provided (6). In the latter stages of the disease, there is clear loss of brain substance, as indicated by gross cortical atrophy and loss of brain weight. Perhaps of equal importance to the total loss of neurons is the atrophy of remaining neurons and their peripheral dendritic processes (7). This atrophy must reduce the total number of cortical synapses (8), and hence, by analogy with computers, reduce the available "computer power."

Positron emission tomography (PET) scanning has demonstrated significant hypometabolism in the parietal cortex of patients with Alzheimer's disease, to an extent that may exceed the pathological changes and true loss of neurons in these

New Directions in Understanding Dementia and Alzheimer's Disease,
Edited by T. Zandi and R. J. Ham, Plenum Press, New York, 1990

31

areas (9). Whether this represents a loss of function from <u>deafferentation</u> in parts of the cortex needed for higher integrative function is unclear. The neuronal changes are concentrated to a greater extent in some areas than others in the brain. For instance, loss of neurons is greatest in the nucleus basalis of Meynert and the septal nuclei, which are responsible for the major part of the cholinergic input to the cerebral cortex (10).

NEUROTRANSMITTER CHANGES

Studies of the content of neurotransmitters, their receptors and the enzymes responsible for their synthesis and degradation have demonstrated a number of alterations. The major changes are in acetylcholine and somatostatin (11,12,13,14). On balance, the evidence indicates that these changes are secondary to loss of the relevant neurons, rather than being the primary cause of the neurological disease. Nevertheless, modulation of neurotransmitter function may offer potential therapy for Alzheimer's disease (15).

NEUROFIBRILLARY TANGLES AND PAIRED HELICAL FILAMENTS

The characteristic argentophilic neurofibrillary tangles (NFTs) within cortical neurons that were first recognized by Alzheimer are composed of masses of paired helical filaments. These react with antibodies against many of the proteins that are present in normal neurofilaments, but are chemically and ultrastructurally different (16,17). They are extremely insoluble, resisting all protein-dissolving agents. In fact, this has formed the basis of methods for their purificatiion (18). NFTs are particularly concentrated in the hippocampus and medial temporal lobes, but also present in neurons throughout the cortex. They are not specific for Alzheimer's disease and are also seen in dementia pugilistica, post-encephalitic Parkinson's disease, and several other conditions. Whether the chemical structure of these NFTs is identical in all the conditions has not yet been demonstrated.

The second pathological hallmark of Alzheimer's disease is the neuritic plaque (NP), which is comprised of swollen nerve elements containing neurofibrillary tangles surrounding a central core of amyloid material. NPs seem particularly to be the hallmark of Alzheimer's disease being directly correlated with the level of dementia (19); more will be said later about the amyloid.

There is a normal accumulation of NFTs and NPs in aging brain, but the rate and extent of accumulation is greater in Alzheimer's disease. Quantitative parameters for diagnosing Alzheimer's disease in different age groups have been presented (20).

ALUMINUM

It was an early recognition in the field of renal dialysis that a few patients developed a progressive dementia, and that this appeared to be associated with accumulation of aluminum in the brain (21,22). Aluminum was also found in higher concentrations in the brains of patients with Alzheimer's disease than in control brains (23). The distribution of the aluminum appears to be somewhat different between the two conditions, and neurofibrillary tangles are not a feature of dialysis dementia. Nevertheless, the role of aluminum in Alzheimer's disease continues to be of great interest (24).

In Alzheimer's disease the aluminum is particularly concentrated in the area of the neurons with NFTs (25,26), a fact which was first demonstrated in the Guamanian-type of ALS-Parkinson's dementia (ALS-PD-dementia) complex that will be discussed later. The administration of aluminum parenterally to experimental animals, particularly the rabbit, produces an accumulation of argentophilic material within neurons which is very similar to that seen in Alzheimer's disease (27). However, these accumulations are straight neurofilaments, with a high concentration of phosphorylated epitopes, which is unusual for neuronal neurofilaments (28); they are not the paired helical filaments associated with Alzheimer's Disease. The rabbits with aluminum encephalopathy show demonstrable functional changes in conditioned reflexes (29). (See Chapter 2.)

CALCIUM

There is evidence of abnormal calcium metabolism in skin fibroblasts from patients with Alzheimer's disease when compared with normal age-matched controls (30). This has suggested that there may be similar abnormalities in the neurons, and that potentially the abnormal intracellular calcium concentration might be responsible for the degenerative events occurring in neurons. As will be discussed later when considering Guamanian ALS-PD-dementia, there have been suggestions that an abnormal calcium-aluminum ratio might underlie the endemic foci of these diseases.

ENDEMIC FOCI OF ALS-PARKINSON'S-DEMENTIA

In New Guinea, Guam, and the Kii Peninsula of Japan, there are foci of high rates of ALS (31). In Guam the patients have the complex of ALS-PD-dementia, which occurs together in many patients, and in isolation in a few. The prevalence of the disease at its peak was more than 100 times that in non-endemic areas. Even within the island of Guam, the high prevalence was to be restricted to some villages and not others. The basis for these endemic foci has been of great concern to all interested in the cause of these diseases. One feature of the island that excited the interest of the neurologists who originally investigated this focus was that the villagers lived on a carbohydrate staple made from the cycad nut. This nut contains a number of different chemicals including cyanide-containing glycosides, and a potent hepatotoxin and carcinogen. The processing of the nut prior to consumption removes most of these substances, but Spencer and his colleagues have recently reactivated the hypothesis that a constituent of the cycad may be the cause of these neurological degenerations (32). They have identified the presence of excitatory aminoacids, particularly B-N-methylamine-L-alanine (BMAA) in cycad, and adduced evidence that these aminoacids can produce neurological degenerations in experimental animals similar to the human diseases. There are still problems to be overcome by this hypothesis. In particular, the very long incubation period is quite unlike that of any known neurotoxin. Nevertheless, there is reason to consider overactivity of excitatory amino acids, especially glutamate, in the etiology of a number of chronic neuronal degenerations (33).

The high concentration of aluminum in the endemic areas has also excited interest, with a suggestion that the streams have low calcium content, with a consequent high aluminum to calcium ratio leading to the disease. More recent evidence has cast doubt on this hypothesis, emphasizing the fact that many streams contain normal amounts of calcium, and that calcium-rich fish make up a large part of the diet (Steele, J., 1988. Personal communication). This matter has once more been raised by the recent finding of the statistical association between prevalence of Alzheimer's disease based upon records of CT scanning units and aluminum contents in the water over the previous 10 years (34), though there are major reservations about the significance of this study (35).

GENETIC FACTORS

In most large series of patients with pre-senile and senile dementia, 5-15% of affected individuals have a familial pattern of disease, in most instances, with dominant inheritance. There have been a number of studies trying to analyze the weight that genetic factors might play in the overall population of patients with Alzheimer's disease, with estimates ranging from 20% to perhaps almost all cases (36,37,38,39,40). The difficulty in the analysis is that 30% to 50% of the population who live to the age of 90 years have Alzheimer's disease, and perhaps an even higher proportion of those who live to an even older age. This can be interpreted in two ways. Either familial aggregation can be a fortuitous event in a high prevalence disease, or genetic factors may play a major role in every case. The high natural frequency of Alzheimer's disease also complicates research studies searching for underlying abnormalities in such patients, since up to 30% of the controls might be expected to have the underlying predilection to develop Alzheimer's disease.

There are clearly reported dominantly inherited early-onset families with Alzheimer's disease in whom genetic factors appear to be absolute (38). There are later onset families in which there may be more question about the inheritance. These families are playing a major role in ongoing research into the underlying cause of Alzheimer's disease.

CHROMOSOME 21

All patients who live beyond the age of 35 years with Down's syndrome (trisomy 21) develop neuropathological changes in the brain that are diagnostic of Alzheimer's disease (41). Most Down's patients living beyond the age of 50 years show a progressive loss of their already somewhat low intellectual function. There is reason, therefore, to consider that chromosome 21 may play some role in Alzheimer's disease (42,43). The precursor of protein for the amyloid which accumulates with Alzheimer's disease is encoded by a gene localized on chromosome 21 (44,45). The gene for some of the early onset familial Alzheimer's disease cases is also localized on chromosome 21. It was originally suggested that triplication of a portion of chromosome 21 containing the amyloid gene might occur in Alzheimer's disease, but again this has proved not to be the case. The inter-relationship between chromosome 21, the amyloid gene and Alzheimer's disease still remains unclear, particularly since some of the later-onset families with Alzheimer's disease have recently been suggested to link to chromosome 14.

DEOXYRIBONUCLEIC ACID

Mann and Yates were the first to notice abnormalities of neuronal nuclei and RNA in neurological degenerations, their first report being in ALS (46). They later reported similar changes in Alzheimer's disease and Parkinson's disease (47,48). They suggested that these changes indicated decreased protein synthesis that might be the final pathway for neuronal degeneration, and that the underlying abnormality might be an alteration in chromatin structure and deoxyribonucleic acid (DNA). Crapper McLachlan and colleagues demonstrated a biochemical abnormality of chromatin structure and digestibility of the DNA by nucleases (49,50,51). Robbins and colleagues, working on xeroderma pigmentosum, recognized that there was a relationship between the neurological degeneration seen in some of these patients and the degree of deficiency of DNA repair for ultraviolet light (52,53). They developed the hypothesis that deficient DNA repair might underlie this and other neurological degenerations. Further work has clearly demonstrated the deficiency of DNA repair, particularly for alkylation damage in skin fibroblasts, peripheral blood lymphocytes, monocytes, and lymphoblasts from sporadic and familial Alzheimer's disease (54,55,56). Whether this repair deficiency is related to a specific enzyme deficit, or to abnormalities of regulatory proteins (DNA binding proteins) (57), is currently not known. Mullaart et al. (58) have recently produced direct evidence of an increased number of DNA breaks or alkali-labile sites in cortical nuclear DNA from patients with Alzheimer's disease compared with controls.

RIBONUCLEIC ACID

Sajdel-Sulkowska et al. (59) confirmed by direct biochemical extraction the observations by Mann et al. (1981) (47) of decreased ribonucleic acid (RNA) in cortical neurons in Alzheimer's disease. They were moreover able to show that the mRNA was particularly reduced, and that the protein synthetic capacity of the mRNA, on a milligram-for-milligram basis, was dramatically reduced compared with control mRNA. All of these observations fitted with the suggestions of Mann and Yates and would be compatible with a hypothesis based on a deficiency of DNA repair and subsequent defective transcription. Sajdel-Sulkowska, et al. (59) also demonstrated a decrease in an RNA's inhibitor in the brains of Alzheimer's disease patients, that was suggested to be responsible for the decreased RNA content, though this observation has not been confirmed.

SUMMARY

There are many theories to explain the cause of Alzheimer's disease. None is mutually exclusive of the others, and it may be that all are correct. The only problem may be that we do not understand which is the primary cause and which are the secondary effects of the primary abnormality in the disease. It is almost certain that Alzheimer's disease, as we recognize it today, is heterogeneous. One has only to think of the early-onset and late-onset familial cases to realize that this is so. All of the theories have experimental evidence to support them, and all have generated experiments to substantiate them. Some of them have generated potential concepts for treatment, none of which at present have proved to be successful. When in the end the underlying etiology of the condition is discovered, it will be possible to fit all of the experimental observations into place.

It appears at present that the most likely breakthroughs in our understanding will come from detailed sequencing of the paired helical filaments and from breakthroughs in the field of molecular genetics studying the gene for familial Alzheimer's disease.

REFERENCES

1. Katzman, R. (1983). Demography, definitions and problems, Ch. 1 in: *The Neurology of Aging.* (eds) Katzman, R. and Terry, R. D., David, F. A., Philadelphia.
2. Plum, F. (1979). Dementia: An approaching epidemic. *Mature, 279*:373-373.
3. Evans, D. A., Funkenstein, H. H., Albert, M. S. (1989). Prevalence of Alzheimer's disease in a community population of older persons. *JAMA, 262*:2551-2556.
4. Bick, K., Amaducci, L., Pepen, G. (1987). *The Early Story of Alzheimer's Disease.* Liviana Press, Padova.
5. Editorial (1987). Alzheimer's disease and associated disorders, Vol 1, pp. 1-4.
6. Hansen, L. A., DeTeresa, R., Davies, D. (1988). Neocortical morphometry, lesion count, and choline acetyltransferase levels in the age spectrum of Alzheimer's disease, *Neurol.* 38:48-54.
7. Schiebel, A. B. and Tomiyasu, V. (1978). Dendritic sprouting in Alzheimer's presenile dementia. *Exper. Neurol.,* 60:1-8.
8. Hamos, J. E., DeGennaro, L. J., Drachman, D. A. (1989). Synaptic loss in Alzheimer's disease and other dementias. *Neurol.,* 39:355-361.
9. Cutler, N. R., Haxby, J. V., Duara, R. (1985). Clinical history, brain metabolism and neuro-psychological function in Alzheimer's disease, *Ann. Neurol.* 18:298-309.
10. Coyle, J. T., Price, D. L., Delong, M. R. (1983). Alzheimer's disease: a disorder of cortical cholinergic innervation. *Science, 219*:1184-1189.

11. Carlson, A. (1983). Changes in neurotransmitter systems in the aging brain and in Alzheimer's disease. Ch 13 in *Alzheimer's Disease*, eds: Reisberg, B., Free Press, New York.
12. Davies, P. (1983). An update on the neurochemistry of Alzheimer's disease, pp. 75-86 in: *Advances in Neurology, 38 The Dementias*. Eds: Mayeux, R. and Rosen, W. G., Raven Press, New York.
13. Francis, P. T., Palmer, A. M., Sims, N. R. (1985). Neurochemical studies of early onset Alzheimer's disease. *New Engl. J. Med., 313*: 7-11.
14. Ferrier, I. N., Cross, A. J., Johnson, J. A. (1983). Neuropeptides in Alzheimer's type dementia. *J. Neurol. Sci., 62*:159-170.
15. Waters, C. (1988). Cognitive enhancing agents: Current status in the treatment of Alzheimer's disease. *Can. J. Neurol. Sci., 15*:249-256.
16. Perry, G., Rizzuto, N., Autilio-Gambetti, L. (1985). Paired helical filaments from Alzheimer's disease patients contain cytoskeletal components. *Proc. Nat. Acad. Sci., 82*:3916-3920.
17. Grundke-Iqbal, I., Johnson, A. B., Wisniewski, H. M. (1979). Evidence that Alzheimer's disease neurofibrillary tangles originate from neurotubules. *Lancet, 1*:578-580.
18. Selkoe, D. J., Abraham, C. R., Podlisny, M. D. (1986). Isolation of low molecular-weight proteins from amyloid plaque fibers in Alzheimer's disease. *J. Neurochem, 46*:1820-1834.
19. Blessed, G., Tomlinson, B. E., Roth, M. (1968). The association between quantitative measures of dementia and of senile changes in the cerebral grey matter of elderly subjects. *Brit. J. Psychiat., 114*:797-811.
20. Khachaturian, Z. S. (1985). Diagnosis of Alzheimer's disease. *Arch. Neurol., 42*:1097-1105.
21. Crapper, D. R., Quittkat, S., Krishnan, S. S. (1980). Intranuclear aluminum content in Alzheimer's disease, dialysis encephalopathy and experimental aluminum encephalopathy. *Acta Neuropath.* (Berl.), *50*:19-24.
22. O'Hare, J. A., Callaghan, N. M. (1983). Dialysisencephalopathy. *Medicine, 62*:129-141.
23. Crapper, D. R., Karlik, S. and De Boni, V. (1978). Aluminum and other metals in senile (Alzheimer's) dementia, pp. 471-484, in Alzheimer's disease: Senile dementia and related disorders. *Aging* (Vol. 7). Eds: Katzman, R., Terry, R. D., Bick, K. L. Raven Press, N.Y.
24. Schwartz, A. S., Frey, J. L., Lukas, R. J. (1988). Risk factors in Alzheimer's disease: Is aluminum hazardous to your health? *Barrow Neurol. Inst. Quart., 4*:2-8.
25. Perl, D. P., Gajdusek, D. C., Garruto, R. M. (1982). Intraneuronal aluminum accumulation in amyotrophic lateral sclerosis and Parkinsonism-dementia of Guam. *Science, 217*:1053-1055.
26. Perl, D. P. (1983). Pathologic association of aluminum in Alzheimer's disease, Ch 15 in: *Alzheimer's Disease*. Ed: Reisberg, B., Free Press, New York.
27. Pendlebury, W. W., Beal, M. F., Kowall, N. W. (1988). Neuropathologic, neurochemical and immunocytochemical characteristics of aluminum-induced neurofilamentous degeneration. *Neurotoxicology, 9*:503-510.
28. Pendlebury, W. W., Munos-Garcia, D., Perl, D. P. (1987). Cytoskeletal pathology in neuro-degenerative diseases. In: (eds) Ehlich, Y. H., Lenox, R. H., Koraecki, E., *Molecular mechanisms of neuronal responsiveness*. Plenum Press, New York, pp. 427-442.
29. Solomon, P. R., Beal, M. F., Pendlebury, W. W. (1988). Age-related description of classical conditioning: A model system approach to memory disorders. *Neuroseology of Aging, 9*:535-546.
30. Peterson, C., Goldman, J. E. (1986). Alterations in calcium content and biochemical processes in cultured skin fibroblasts from aged and Alzheimer donors. *Proc. Nat. Acad. Sci., 83*:2758-2762.

31. Tsubaki, T., Yase, Y. (eds) (1988). Amyotrophic lateral sclerosis, Excerpta Medica Int. Conf. Series 769, Amsterdam.

32. Spencer, P. S., Palmer, V., Ohta, M. (1988). Cycad, a suspect etiological factor for Guam ALS/PD, is associated with motor neuron disease in Irian Jaya, Indonesia and Kii peninsula, Japan, pp. 35-40. In: Tsubaki, T. and Yase, Y. (eds): Amyotrophic lateral sclerosis, Excerpta Medica Int. Conf. Series 769, Amsterdam.

33. Plaitakis, A., Caroscio, J. T. (1987). Abnormal glutamate metabolism in amyotrophic lateral sclerosis. *Neurol., 22*:575-579.

34. Martyn, C. N., Osmond, C., Wardson, J. A. (1989). Geographical relation between Alzheimer's disease and aluminum in drinking water. *Lancet,* 59-62.

35. Editorial, *The Lancet*, January 14, 1989. Aluminum and Alzheimer's disease, 82-83.

36. Heston, L. L., Mastri, A. R., Anderson, V. E. (1981). Dementia of the Alzheimer's type. *Arch. Gen. Psych., 38*:1085-1090.

37. Heston, L. L. (1983). Genetic studies of dementia, Ch 6 in *The Epidemiology of Dementia*. Eds: Mortimer, J. A., Schuman, L. M., Oxford Univ. Press, New York, 101-114.

38. Nee, L. E., Polinsky, R. J., Eldridge, R. (1983). A family with histologically confirmed Alzheimer's disease. *Arch. Neurol., 40*:203-208.

39. Matsuyama, S. S. (1983). Genetic factors in dementia of the Alzheimer-type. Ch 21 in *Alzheimer's Disease*. Ed: Reisberg, B., Free Press, New York, 155-160.

40. Farrer, L. A., O'Sullivan, D. M., Cupples, L. A. (1989). Assessment of genetic risk for Alzheimer's disease among first degree relatives. *Ann. Neurol., 25*:485-493.

41. Wisniewski, K. E., Wiskiewski, H. M., Wen, G. Y. (1985). Occurrence of neuropathological changes and dementia of Alzheimer's disease in Down's syndrome. *Ann. Neurol., 17*:278-282.

42. Editorial (1987). Alzheimer's Disease, Down's syndrome and Chromosome 21. *Lancet, 1*:1011-1012.

43. Hardy, J. (1988). Molecular biology and Alzheimer's disease. *TINS, 11*: 293-294.

44. Tanzi, R. E., Gusella, J. F., Watkins, P.C. (1987). Amyloid B protein gene: cDNA, mRNA distribution and genetic linkage near the Alzheimer's locus. *Science, 235*:880-884.

45. St. George-Hyslop, P. H., Tanzi, R. E., Polinski, R. J. (1987). The genetic defect causing familial Alzheimer's disease maps to chromosome 21. *Science, 235*:885-890.

46. Mann, D. M. A., Yates, P. O. (1974). Motor neuron disease. *J. Neurol. Neurosurg. Psychiat., 37*:1036-1047.

47. Mann, D. M. A., Neary, D., Yates, P.O. (1981). Alterations in the protein synthetic capability of nerve cells in Alzheimer's disease. *J. Neurosurg. Psychiat., 44*:97-105.

48. Mann, D. M. A., Yates, P.O. (1983). Pathological basis of neuro-transmitter changes in Parkinson's disease. *Neuropath. Exp. Neurobiol., 8*:3-14.

49. Lewis, P. N., Lukin, W. J., DeBoni, V. (1981). Changes in chromatin structure associated with Alzheimer's disease. *J. Neurochem., 37*:1193-1202.

50. Crapper-McLachlan, D. R., Lewis, P. N., Lukiw, W. J. (1984). Chromatin structure in dementia. *Ann. Neurol., 15*:329-334.

51. Crapper-McLachlan, D. R., Lewis, P. N. (1985). Alzheimer's disease: errors in gene expression. *Canadian J. Neurol. Sci., 12*:1-5.

52. Robbins, J. H. (1978). ICN-UCLA Symposia on molecular and cellular biology, Vol IX, Academie Press, 603-629.

53. Andrews, A. D., Carrett, S. R., Robbins, J. H. (1978). Xeroderna pigmentosum: Neurological abnormalities correlate with colony forming ability after UV-radiation. *Proc. Nat. Acad. Sci., 75*:1984-1988.

54. Robbins, J. H., Otswka, F., Tarone, R. E. (1983). Radiosensitivity in Alzheimer's disease and Parkinson's disease. *Lancet, 1*:468-469.
55. Robinson, S. H., Munzer, J. S., Tandan, R. (1987). Alzheimer's disease cells have defective repair of alkylating agent-induced DNA damage. *Ann. Neurol., 21*:250-258.
56. Tandan, R., Robison, S. H., Bartlett, J. D. (1988). DNA damage and repair in ALS and Alzheimer's disease lymphoid cells and monocytes. In: *Amyotrophic Lateral Sclerosis.* Eds: Tsubaki, T., Yase, Y., Elsevier, New York, 113-118.
57. Wetmur, J. G., Casals, J., Elizan, T. S. (1984). DNA binding protein, profiles in Alzheimer's disease. *J. Neurol. Sci., 66*:201-208.
58. Mullaart, E., Boerrigter, M. E. T., Ravid, R. (1989). Increased levels of DNA breaks in cerebral corte of Alzheimer's disease patients. In press.
59. Sajdel-Sulkowska, E. M., Marotta, C. A. (1984). Alzheimer's disease brain: Alterations in RNA levels and in a ribonuclease-inhibitor complex, *Science, 225*:947-949.

THE REVERSIBLE DEMENTIAS: DO THEY REVERSE?

A. Mark Clarfield, MD, CCFP, FRCPC

Chief, Division of Geriatrics
Sir Mortimer B. Davis - Jewish General Hospital
Associate Professor and Assistant Dean
McGill University
Montreal, Canada

INTRODUCTION

Over the last two decades, the possibility of finding (and curing) the reversible dementias has gained wide currency. As a result, multiple bodies (1-4) and individuals (5-9) have recommended more or less extensive workups for the demented patient, hoping thereby to ensure that no potentially reversible cases are missed.

More recently, some dissent has been expressed, at least with respect to the probable incidence of such reversibility. In sum, the arguments presented go as follows: Because of various biases in the literature, the incidence of potential reversibility has been overestimated. These biases primarily involve those of selection, as well as referral filter (10). For example, many studies have not reported the age structure of their study sample. Others examined subjects who were substantially younger than most of those actually suffering with dementia, an unfortunate practice followed for many other conditions primarily affecting the elderly (11). As well, most studies emanated from tertiary care referral centres, and many will have examined an atypical subset of demented patients. In the chapter that follows, a critical review of the literature suggests that, in fact, the incidence of reversibility is not higher than 11% and, in fact, is probably much lower. A recent community-based study has in part corroborated these findings (12).

Given the fact that the incidence of reversible dementias is probably far lower than had previously thought to be the case, what of the authoritative recommendations published with respect to the laboratory workup for these patients? (1-4) Common sense suggests that the higher the actual incidence of reversibility, the more comprehensive should be the "routine" battery of investigations. On the other hand, as we now know that reversibility is not nearly as frequent as was previously thought to be the case, there might be reason to question the validity of some of the previous recommendations.

Given that there are so many different conditions which can cause dementia, and that the literature is not yet clear on the cost-effectiveness of every test required for a "rule out reversible" approach, a Canadian consensus conference

New Directions in Understanding Dementia and Alzheimer's Disease,
Edited by T. Zandi and R. J. Ham, Plenum Press, New York, 1990

was organized to try to provide reasonable guidelines for the primary care clinician. Although an American consensus conference on the same subject has been held as recently as 1986 (3), it took place before the recent critical reviews appeared in print (6,8,9).

As well, a Canadian conference was felt to be necessary for the following reasons:

1. There are some significant disagreements between the American (3) and British (4) sets of recommendations.

2. The Canadian health care system differs significantly from its British and American counterparts. For example, specific differences between the Canadian and American systems involve, among others, the following issues:

- the extent of the practice of defensive medicine and differing malpractice climate
- the difference in profile and strength of the respective medical/industrial complexes
- diagnostic entrepreneurial differences
- regionalisation of technology and government role in decision making (e.g., the availability of CT scan)
- relative importance of primary care system in Canada

It was thus felt that a Canadian set of guidelines was needed, taking into account the strengths, weaknesses and political/social realities of our own health care system.

3. Given the strong role of the Family Physician in Canada's health care system, especially with respect to the care of the elderly (13), it was felt that guidelines should be drawn up for Canada's primary care physicians considered in the case of dementia to the family doctor, general internist and general psychiatrist. While other sets of recommendations (1-4) did not specify their medical audience, it was implied that these guidelines were addressed primarily to specialists in the field of dementia.

4. In Canada, there is an increasing preoccupation with the appropriate use of medical technology. One of the issues involved in the workup of dementia is concerned with the use of neuroimaging procedures. As in the United Kingdom and unlike that which pertains in the United States, most Canadian primary care physicians do not have direct access to such technology. (We do not address here that thorny issue of whether such access should be broadened.) (14,15).

5. Enough Canadian physicians have expressed disagreement with the recommendations of previous bodies that we felt it was timely to re-examine the issue.

6. Recently, there has been some unhappiness with the view that the assessment of dementia should be considered merely a dichotomy between reversible and non-reversible causes (5), that the physician can "wash his/her hands" of the case if no reversibility is found.

7. Many ethnic groups are represented among Canada's elderly. It was felt timely to provide some guidance with respect to the problem of accurate diagnosis of the dementia states in these groups.

8. In order to provide Canadian scientists with a research agenda for the dementias.

Over a period of two days and after months of preparatory work, a meeting involving 34 experts from Canada, as well as four from the United States took place in Montreal, Canada. During this meeting, the issues relating to the diagnosis and assessment of dementia were thoroughly debated and a consensus statement drafted. Publication of these findings is planned for the near future in the Canadian Medical Association Journal. (Those who would like a copy of the full report may write to the author of this chapter.)

There are some patients who will turn out to have a potentially reversible cause for their dementia, and even fewer who actually do improve with therapy. Nevertheless, each demented patient should be carefully assessed, involving a complete history, a guided physical and the appropriate laboratory tests. If this assessment is accomplished in a competent manner, very few reversible cases will be missed; at the same time, very few false positives will have been generated (16). In the end, neither too much nor too little will have been effected, with a minimum generation of iatrogenic disease and/or false hope for demented patients and their families.

*Although it has been known since at least 1888 that dementia could be caused by organic illness (1), the idea that it could be reversed first came with the successful treatment of neurosyphilis in the 1940s (2). There was little further progress until 1965 when Adams and associates (3) reported the reversal of dementia secondary to normal pressure hydrocephalus after doing a cerebrospinal fluid shunt. As it became clear that many conditions (4), ranging from atrial myxoma (5) to zoonoses (6), could cause dementia, studies were done to establish the relative frequency of these conditions (7-38). On the basis of reviews of some of these studies, various authors (23, 39-42) have claimed that between 10% and 40% of dementias might be reversible.

After this work, much has been written about the appropriate workup of dementia, with particular attention to the search for reversible causes. One can distinguish between at least two schools of thought regarding the extensiveness of workup. One philosophy, often espoused by British or Canadian authorities (43-47), holds that truly reversible causes of dementia present infrequently, and recommends that the workup should be more clinically oriented and relatively restricted with respect to the routine use of laboratory and radiologic facilities. American authorities have a more technologically oriented approach to medicine (48, 49), and recommend an elaborate workup on nearly all demented patients (40, 42, 50-53).

From the point of view of cost-effectiveness and common sense, the higher the true incidence of reversible dementias, the more extensive and searching the routine investigation of the demented population should be. However, if reversible causes are relatively rare, one could more safely recommend a less radical diagnostic approach for most patients.

The prevalence of dementia is rising rapidly due to increasing numbers of elderly and very elderly people in the world (54). The costs (both financial and general) of caring for these people are high for the family, the health system, and society as a whole (55, 56). Thus, it would be of interest both to clinicians and health planners to know the true incidence of reversible dementias. To determine this figure (and its breakdowns into the various conditions), a critical review of the literature of studies addressing this question was done.

METHODS

A computer search for 1966 to 1987 using Index Medicus was done (keywords: dementia and etiology, or diagnosis, or differential diagnosis, or follow-up studies, or assessment, or treatable dementia, or reversible dementia). In addition, careful follow-up of all bibliographic hints in the articles generated by the computer search was done. Standard textbooks of psychiatry, neurology, and geriatics were consulted for further references. Only clinical studies (7-38) were included because autopsy research (57-62) tends to emphasize structural changes in

*See page 53.

Table 1. Sample Size, Patient Age, Sex, and Sample Provenance Data from 32
Studies of Patients with Dementia Used for Analysis

Reference Year (number)	Patients	Age*	Number of Women	Origin of Patient
	n	yrs	%	
Marsden and Harrison, 1972 (7)	106	Presenile	...	Neurology department; hospital
Pearce and Miller, 1973 (8)	63	80% presenile	...	...
Fox et al., 1975 (9)	40	75 (63-93)	48	Neurology department; hospital
Freemon, 1976 (10)	60	66.2	0	Veterans hospital
Harrison and Marsden, 1977 (11)	43	Presenile	...	General hospital
Katzman, 1977 (12)	56	...	...	...
Victoratos et al., 1977 (13)	52	(40-73)	56	Neurology department; hospital
Kokmen et al., 1980 (14)	102	...	...	Community
Reifler and Eisdorfer, 1980 (15)	82	77	68	Geriatric outpatients department
Barnes and Raskind, 1981 (16)	64	78 (63-101)	72	Nursing home
Garcia et al., 1981 (17)	94	71 ± 10	...	Neurology department; rehabilitation hospital
Hutton, 1981 (18, 19)	100	67 ± 11.7†	...‡	Veterans hospital
Rabins, 1981 (20)	41	...	...	Psychiatry hospital
Smith and Kiloh, 1981 (21)	200	57.7 (23-83)	49	Neuropsychiatry hospital
Benson and Cummings, 1982 (22)	90	...	...‡	Neurology department; Veterans hospital
Delaney, 1982 (23)	100	56 (43-80)	32	Neurology department; Veterans hospital
Freemon and Rudd, 1982 (24)	110	66.5†	0	Veterans hospital
Maletta et al., 1982 (25)	100	...§	...‖	Geriatric outpatients department;¶ Veterans hospital
Sabin et al., 1982 (26)	111	...**	...	Nursing home
Martin et al., 1983 (27)	63	...‡	62	Psychiatry hospital
Larson et al., 1984 (28)	107	75.8 ± 8.1	71	Geriatric outpatients department
Roca et al., 1984 (29)	65	...	...	General hospital
Folstein et al., 1985 (30)	36	...††	44	Community
Gilchrist et al., 1985 (31)	24	73 ± 6.8†	60	Psychiatry hospital
Hammerstrom and Zimmer, 1985 (32)	80	...‡‡	...	Psychiatry department; hospital
Larson et al., 1985 (33)	200	75.7 ± 7.5	...	Geriatric outpatients department
Renvoize et al., 1985 (34)	150	78 (59-94)	67	Psychiatry department; hospital
Schoenberg et al., 1985 (35)	80	78.6	65	Community and hospital
Erkinjuntti et al., 1986 (36)	152	79.2 ± 7.3	66	Internal medicine department; general hospital
Sayetta, 1986 (37)	41	79	0	Community
Oppenheim, 1987 (personal communication)	83	69.9	55	Psychiatry outpatients department
Pfeffer et al., 1987 (38)	186	77.5 ± 6.8	48	Community
Totals				
Mean ± SE	...	72.3 ± 1.7	47.9	...
Median	82.5	...	...	...
Range	24-200	...	...	...

* Age is given as mean; mean (range); (range); or mean ± SD, depending on information available.
† Estimate from data provided.
‡ Probably mostly men.
§ Majority 60+.
‖ "Majority male."
¶ Most patients referred from neurology service.
** All but two ♀.
†† 65+.
‡‡ 55+.

the brain (for example, senile dementia of the Alzheimer type compared with multi-farct dementia) and could not supply much data concerning the potential reversibility of conditions. In the spirit of metanalysis and to minimize the effects of publication bias, a letter to the editor, one published and one unpublished personal communication, and two abstracts were also included in the analysis.

In the first analysis, the size of the study sample, age of patients studied, and type of facility in which the study took place were noted. Next, a collation of the differential diagnoses for the dementia syndrome was made. The conditions were classified as "potentially reversible" (defined as infection, metabolic, neoplasm, normal pressure hydrocephalus, subdural hematoma, depression, and drugs) or "irreversible" (Alzheimer's disease, multi-infarct dementia, alcohol, Parkinson's disease, Huntington's disease, post-anoxic, post-trauma). Neither multi-infarct dementia or dementia related to alcohol were included among the potentially reversible causes; the reasons for these decisions will be addressed below. In some cases, the authors provided a figure for potentially reversible causes without specifying the conditions to which they were referring. In these cases, the authors' figures were accepted. In papers where the data were supplied, the percentage of "partly reversed" or "fully reversed" dementias was calculated. Although not all authors clearly defined partial or full reversal, for the purpose of analysis, cases in which there was doubt were placed in the fully reversed category.

RESULTS

Altogether, 32 studies reported between 1972 and 1987 were analyzed (Table 1). The median size of study sample was 82.5, and 2889 patients were involved. The mean age of patients was 72.3. Almost half of the studies (14 of 32) provided no mean age for their study samples. Of the 18 studies that reported the ages of their subjects, one half examined subjects with a mean age of less than 70 years. In 13 studies that provided follow-up data on reversibility, the mean age of subjects (68.9 years) was lower than the average subject age for all studies surveyed. Most studies (76.6%) originated in tertiary care centers; more than three quarters of studies used in-patient popula tions. Two (6.6%) studies took place in nursing homes and four (13.3%) involved primarily community-based research.

An overview of the diagnoses causing dementia is shown in Table 2. Alzheimer's disease was the commonest cause (56.8%), followed by multi-infarct dementia (13.3%), depression (4.5%), and alcohol (4.2%). Table 3 shows a breakdown of reversible causes of dementia indicating that 13.2% were potentially reversible. In almost two thirds of the studies (19 of 30), no information was provided about the follow-up of potentially reversible cases; that is, whether the potentially reversible cases discovered in these 19 studies actually improved with treatment (either partially or fully) could not be determined.

For those 11 studies in which follow-up data were provided (a total of 1051 patients), 8% of patients reversed partially and 3% completely (total, 11%). The quality of follow-up varied, ranging from retrospective chart review by one person to careful prospective consensus committee investigation of patients followed months and even years after initial diagnosis. Altogether, the 11 studies that provided follow-up specifically described 103 patients with dementia that either partially or fully recovered (Table 4). In nearly one third of these patients, drugs were the cause of the dementia (28.2%), followed by depression (26.2%) and metabolic conditions (15.5%).

DISCUSSION

Is it important to know the true incidence of reversible dementias? Clearly, we cannot afford the "million dollar workup" for every person with global impairment of cognition; yet there are many causes of dementia. Haase (4) lists more than 60 disorders; the list continues to grow, with human immunodeficiency virus being the latest addition (63). Many of these diseases are considered potentially reversible (the National Institute of Aging [42] lists 41 in its 1980 report), and it would be tragic if truly reversible cases were frequently missed.

Diagnosis and Follow-up

Thus, a determination of the true incidence of reversible dementias should help physicians make general diagnostic recommendations. The more common true (not potential) reversibility is, the more extensive the workup should be for all demented patients. If reversibility turns out to be rare, a more conservative approach should be followed.

Before proceeding, it should be pointed out that diagnosis of the dementia is based on clinical findings. Laboratory tests follow to classify according to cause.

Table 2. Breakdown of Causes of Dementia in 32 Studies

Study	Alzheimer Type	Multi-Infarct	Mixed	Infection	Metabolic	Neoplasm	Normal Pressure Hydrocephalus	Subdural Hematoma
				n(%)				
Marsden and Harrison	48(45.3)	8(7.5)	...	4(3.8)	...	8(7.5)	5(4.7)	...
Pearce and Miller	58(92.1)	...	...	1(1.6)	1(1.6)	2(3.2)	...	1(1.6)
Fox. *et al.*	35(87.5)	...	...	...	3(7.5)	1(2.5)	1(2.5)	...
Freemon	26(43.3)	5(8.3)	...	...	...	2(3.3)	7(11.7)	...
Harrison and Marsden	19(44.2)	5(11.6)	...	...	...	2(4.7)	3(7.0)	...
Katzman	39(69.6)	4(7.1)	...	...	2(3.6)	1(1.8)	2(3.6)	...
Victoratos *et al.*	30(57.7)	5(9.6)	...	2(3.8)	...	4(7.7)	1(1.9)	1(1.9)
Kokmen *et al.*	34(33.3)	18(17.6)	12(11.8)	1(1.0)	1(1.0)	2(2.0)	...	2(2.0)
Reifler and Eisdorfer	48(58.5)	...	...	...	...	...	...	...
Barnes and Raskind	36(56.3)	17(26.6)	...	...	...	...	...	...
Garcia *et al.*	39(41.1)	8(8.4)	7(7.4)	...	...	1(1.1)	1(1.1)	...
Hutton	22(22)	12(12)	...	2(2)	7(7)	4(4)	1(1)	...
Rabins	29(70.7)	8(19.5)	...	...	2(4.9)	...	1(2.4)	1(2.4)
Smith and Kiloh	84(42)	22(11)	...	1(0.5)	2(1)	3(1.5)	8(4)	...
Benson and Cummings	22(24.4)	31(34.4)	...	...	...	2(2.2)	4(4.4)	...
Delaney	49(49)	22(22)	...	3(3)	2(2)	5(5)	2(2)	3(3)
Freemon and Rudd	76(69.1)	8(7.3)	...	1(0.9)	2(1.8)	1(0.9)	5(4.5)	3(2.7)
Maletta *et al.*	43(43)	10(10)	...	...	1(1)	1(1)	1(1)	...
Sabin *et al.*	85(76.6)	...	...	...	...	...	...	...
Martin *et al.*	54(85.7)	6(9.5)	...	...	1(1.6)	...	...	...
Larson *et al.*	74(64.3)	4(3.5)	...	...	4(3.5)	...	...	2(1.7)
Roca *et al.*	14(21.5)	23(35.4)	...	...	5(7.7)	...	2(3.1)	...
Folstein *et al.*	12(33.3)	6(16.7)	3(8.3)	...	...	...	...	...
Gilchrist *et al.*	18(75.0)	1(4.2)	...	...	...	...	...	...
Hammerstrom and Zimmer	34(42.5)	10(12.5)	...	1(1.3)	...	...	...	...
Larson *et al.*	139(69.5)	2(1.0)	...	...	8(4.0)	...	...	...
Renvoize *et al.*	143(95.3)	...	...	2(1.3)	1(0.7)	1(0.7)	...	...
Schoenberg *et al.*	44(55)	11(13.8)	...	2(2.5)	...	...	1(1.3)	...
Erkinjuntti *et al.*	35(23)	110(72.4)	...	...	1(0.7)	1(0.7)	1(0.7)	...
Sayetta	27(65.9)	13(31.7)	...	...	...	...	...	...
Oppenheim	64(77.1)	4(4.8)	...	...	...	...	...	...
Pfeffer *et al.*	162(87.1)	10(5.4)	...	...	...	1(0.5)	...	...
Total	1642(56.8)	383(13.3)	22(0.8)	18(0.6)	44(1.5)	42(1.5)	47(1.6)	13(0.4)

Yet there remains much confusion as to how dementia is diagnosed. The Diagnostic and Statistical Manual of Mental Disorders III (DSM III) remains the classical reference (64), yet clinicians will often use less exacting criteria.

QUALITY OF THE DATA

This survey shows several problems that are inherent in any analysis of many studies. First, different authors used different systems of diagnostic classification that changed over the 15 years separating the first and last studies. This practice made it difficult to consistently compare figures among studies. Some works included depression as a cause of dementia, whereas others differentiated it from dementia. Many studies did not always carefully define their terms, and thus one is lead to wonder whether cases of delirium (acute or subacute) might have been included within the roster of supposedly demented patients. For example, one study (9) included a patient whose symptoms had been present for only 1 month, suggesting that in this case, use of the term "dementia" might have been inappropriate. Use of terms such as "unknown," "other," and "miscellaneous" may have obscured causes that could unfortunately not be further analyzed. However, overall these cases only made up 6.9% of the total.

Length of follow-up varied: Sometimes it was not specified (9) and sometimes a fairly long period (2 years) was reported (28, 33). Studies differed from each other in the form and completeness of the diagnostic workup. Even within studies, not all subjects had the same examinations. For example, in one study (16) only some subjects had computed tomography (CT) scans and determination of B12 levels. More importantly, study sites included various settings: tertiary referral centers, general hospitals, nursing homes, various kinds

Table 2. (Continued)

Depression	Drugs	Post Trauma	Anoxic	Huntington Disease	Parkinson Disease	Alcohol	Miscellaneous	Not Demented	Total
					n(%)				
8(7.5)	2(1.9)	1(0.9)	...	3(2.8)	...	6(5.7)	13(12.3)	[15]*	106
...	...	...	...	...	...	...	...	...	63
...	...	...	...	...	...	...	...	...	40
1(1.7)	5(8.3)	...	...	4(6.7)	...	4(6.7)	6(10)	...	60
3(7.0)	2(4.7)	...	...	...	...	4(9.3)	5(11.6)	...	43
...	...	...	...	3(5.4)	1(1.8)	3(5.4)	1(1.8)	...	56
...	...	1(1.9)	1(1.9)	...	1(1.9)	1(1.9)	3(5.8)	2(3.8)	52
...	...	1(1.0)	...	...	6(5.9)	4(3.9)	21(20.6)	...	102
18(22.0)	...	...	...	...	...	...	2(2.4)	14(17.1)	82
[1]	...	3(4.7)	1(1.6)	1(1.6)	1(1.6)	1(1.6)	4(6.3)	[6]	64
15(15.8)	...	...	3(3.2)	...	3(3.2)	...	1(1.1)	16(16.8)	94
14(14)	2(2)	2(2)	1(1)	3(3)	2(2)	12(12)	8(8)	8(8)	100
...	...	...	...	...	...	...	...	...	41
10(5)	...	5(2.5)	1(0.5)	5(2.5)	...	30(15)	3(1.5)	26(13)	200
6(6.7)	...	...	...	...	5(5.6)	...	18(20)	2(2.2)	90
Excluded	4(4)	...	...	1(1)	4(4)	3(3)	2(2)	...	100
3(2.7)	2(1.8)	...	...	1(0.9)	...	8(7.3)	...	...	110
24(24)	1(1)	...	...	2(2)	...	7(7)	4(4)	6(6)	100
...	...	...	...	...	...	...	26(23.4)	...	111
...	...	...	...	...	...	...	2(3.2)	...	63
1(0.9)	6(5.2)	...	...	...	[7]	4(3.5)	5(4.3)	15(13.0)	115
...	...	...	...	...	2(3.1)	6(9.2)	13(20)	...	65
...	...	...	...	...	...	...	15(41.7)	...	36
...	...	...	...	...	...	5(20.8)	...	...	24
11(13.8)	...	...	...	...	1(1.3)	12(15)	4(5)	7(8.8)	80
10(5.0)	10(5.0)	...	...	...	[10]	8(4.0)	15(7.5)	8(4.0)	200
1(0.7)	...	...	...	...	1(0.7)	1(0.7)	...	...	150
...	...	...	...	...	...	2(2.5)	20(25)	...	80
...	...	...	...	...	4(2.6)	...	...	...	152
...	1(2.4)	...	...	...	...	...	...	...	41
6(7.2)	...	...	...	...	2(2.4)	...	3(3.6)	4(4.8)	83
...	6(3.2)	...	...	2(1.1)	2(1.1)	...	3(1.6)	...	186
130(4.5)	42(1.5)	13(0.4)	7(0.2)	25(0.9)	34(1.2)	121(4.2)	198(6.9)	108(3.7)	2889(100)

* Brackets indicate that this number was not included in the totals.

of clinics, as well as community-based studies. This variety added to the difficulty in comparing results.

Some studies, although methodologically sound, could not include enough subjects needed to show positive results for reversible dementias. For example, in the community-based research of Folstein and co-workers (30) the authors administered (among other things) a mini-mental status test to 3481 adults chosen through a probability sample in Eastern Baltimore. Forty-four subjects were found to be demented after confirmatory examination by psychiatrists; 36 of these patients consented to a full dementia workup. However, no cases of reversible dementia were discovered. Because reversible cases occur relatively infrequently, this negative finding does not mean that reversible cases do not exist in the community, but rather that it would be unlikely for many of them to be picked up in such a small sample. The other four partly or fully community-based studies (14, 35, 37, 38) discovered several cases of potentially reversible disease but not described any follow-up.

One must be wary of overdiagnosing dementia because the diagnosis itself is not foolproof. In one study (65), in only 15 of 35 patients diagnosed as having presenile dementia, did the diagnosis hold on follow-up years later. Ron and colleagues (66) reported similar results in another center.

Wilson and associates (67) examined some of the same papers surveyed in this study and found serious deficiencies in methodology in almost all papers surveyed. Initially, Wilson's group examined 34 articles; however, preliminary review eliminated all but 11 (7 of which were also examined in my analysis) (7, 9, 13, 21, 23, 24, 28). The remaining 4 were not included in my review for various reasons: 2 papers (66, 68) did not supply enough information to enable inclusion in

Table 3. Patients with Potentially Reversible, Partly Reversed, and Fully Reversed Dementia

Study	Number of Patients			
	In Study	Potentially Reversible Disease	Partially Reversed Disease	Fully Reversed Disease
	n	←	*n (%)*	→
Marsden and Harrison	106	27(25.5)	. . .	. . .
Pearce and Miller	63	6(9.5)	. . .	. . .
Fox. *et al.*	40	5(12.5)	1(2.5)	2(5.0)
Freemon	60	18(30)	7(11.6)	5(8.3)
Harrison and Marsden	43	14(32.5)	. . .	. . .
Katzman	56	5(8.9)	. . .	. . .
Victoratos *et al.*	52	8(15.4)	3(5.8)	0
Kokmen *et al.*	102	6(5.9)	. . .	. . .
Reifler and Eisdorfer	82	18(22.0)	. . .	. . .
Barnes and Raskind	64	0	. . .	. . .
Garcia *et al.*	94	17(18.1)	. . .	. . .
Hutton	100	30(30)	10(10)	8(8)
Rabins	41	4(9.8)	. . .	. . .
Smith and Kiloh	200	24(12)	14(7.0)	3(1.5)
Benson and Cummings	90	12(13.3)	. . .	. . .
Delaney	100	23(23)	23(23)*	
Freemon and Rudd	110	16(14.5)	7(6.4)	5(4.5)
Maletta. *et al.*	100	28(28)	. . .	. . .
Sabin. *et al.*	111	26(23.4)	. . .	. . .
Martin *et al.*	63	1(1.6)	0	0
Larson *et al.* (1984)	107	15(14)	13(12.1)	3(2.8)
Roca *et al.*	65	. . .	. . .	. . .
Folstein *et al.*	36	0	0	0
Gilchrist *et al.*	24	0	. . .	. . .
Hammerstrom and Zimmer	80	12(15)	. . .	. . .
Larson *et al.* (1985)	200	28(14)†	28(14)	2(1.0)
Renvoize *et al.*	150	7(4.7)	5(3.3)*	
Schoenberg *et al.*	80	5(6.3)	. . .	. . .
Erkinjuntii. *et al.*	152	3(2.0)	. . .	. . .
Sayetta	41	1(2.4)	. . .	. . .
Oppenheim	83	6(7.2)	1(1.2)	4(4.8)
Pfeffer *et al.*	186	7(3.8)	. . .	. . .
Mean	90‡	12(13.2)§	7.6(8.0)‖	2.9(3.0)‖

* Could not distinguish between partially and fully reversed; data not included in totals.
† Potentially reversible cases (28) less than total of partially and fully reversed (30) because some conditions reversed that originally had been considered irreversible.
‡ Figures based on 32 studies reporting.
§ Figures based on 31 studies reporting.
‖ Figures based on 11 studies reporting.

my tables; one paper examined all psychiatric diagnoses rather than just dementia (69); and one paper (70) included data that were also included in a later publication by the same authors (21), which I did include in my survey.

Even after excluding 23 of the initial 34 articles surveyed (the excluded papers were not referenced by Wilson and colleagues [67], and so it is not possible to determine whether there was any overlap with the studies examined in my review), these authors found that "most studies did not approach a reasonable standard for definitive information on natural history." Specifically, three studies met none of the six criteria used to judge a natural history study; five others fully or partially met only one criterion. One study, by Larson and associates (28), met all but one criterion. As Wilson and colleagues (67) point out, this study was also the only one that was based on an inception cohort, the most basic requirement for a study of natural history.

Of the 11 studies (Table 3) that reported reversed dementia, 5 were analyzed by Wilson and colleagues (67). Interestingly, the paper by Larson and associates (28) that was judged so favorably in Wilson's analysis, reported figures of fully (2.8%) and partly (12.1%) reversed disease not dissimilar to the overall figures shown in Table 3 (3% and 8% respectively) of my analysis.

THE TRUE INCIDENCE OF REVERSIBLE DEMENTIAS

Table 3 shows that from the 11 studies that reported adequate follow-up data, 11% of the cases of dementia reversed partly (8%) or fully (3%). This figure of 11% falls at the low end of the range usually supplied for the proportion of dementias thought to be reversible. Yet, there are indications that even this number may represent at best an inaccurate and at worst a significant overestimate of the true incidence of reversibility.

First, the quality of age data supplied as well as the actual overall mean age (72.3 years) make it difficult to generalize these studies to the demented population with which most clinicians deal. It is well known that dementia is primarily a problem of the elderly and particularly the very elderly (71). In a community survey, Sayetta (37) found the prevalence of dementia to rise geometrically from zero at age 60 to greater than 50% at age 95. Yet, as shown in Table 1, not all studies properly reported the age structure of their samples. In studies that reported age data, one half examined relatively young populations (less than 70 years). Almost one half of the studies reported no mean age, which is regrettable not only for studies of dementia, but for other clinical studies relevant to the elderly as well (72).

The finding in one of the studies reviewed (21) that investigations for potentially reversible disease were more rewarding in younger patients is of interest in this context. Specifically, 20.7% of patients aged less than 65 had potentially reversible disease compared with only 5.4% of patients greater than 65 years, where most dementias are to be found. On the other hand, Hutton (18, 19) found the opposite; his "improved" patients tended to be slightly older (mean, 71.3 + 7.9 years) compared with the "unimproved" cases (mean, 66.5 + 12.6 years). Delaney (23) found that age did not differentiate potentially treatable from idiopathic causes.

Although the number of improved cases described in these three papers is too small to enable one to draw firm conclusions, it is interesting to note that normal pressure hydrocephalus was the most frequently reversed cause of dementia in younger patients (21); metabolic, tumor, and drugs were the most frequently reversed causes in older patients (18, 19). Depression was equally represented among the young and old.

Almost all studies originated from tertiary care centers, possibly biasing the cases towards rarer and more diagnostically subtle conditions. For example, one study (24) actually generated its patients from a retrospective review of the clinical charts of patients who had had a CT scan for workup of cognitive decline (the authors did recognize this bias in their discussion). Another study (36) evaluated all demented patients from among 2000 consecutive admissions to a large university teaching hospital department of medicine. In a third study (7), all patients had been seen by either a psychiatrist or a neurologist or both, before admission for the dementia workup described. Similarly, Maletta and associates (25) examined patients referred to a clinic from a neurology service.

In support of the contention that the literature probably overestimates the true incidence of reversibility, the one community-based study (37) that did describe follow-up of its potentially reversible causes reported only one, with no truly reversed cases found. The other four community-based studies (30, 35, 37, 38) taken as a group showed an average of only 4.6% potentially reversible causes (with no further follow-up reported). This figure is much lower than the overall rate of 13.2% computed for all studies examined (Table 3). This large discrepancy hints at the possibility of an important selection bias.

Sackett and coworkers (73) ask the following question about whether medical research can be generalized: "Is the site [of the study] so dissimilar to your own that its results, even if they are valid would not apply in your practice?"

Several other authors (39, 43, 46, 47, 74, 75) have published opinions and data that support the contention that selection bias is a serious impediment to the determination of the true incidence of reversible dementias.

An analogous phenomenon occurred when the concept of secondary hypertension first became known (73). Physicians thought they were faced with a high prevalence of a devastating condition, a significant proportion of which might be potentially reversible. It followed that there might be many potentially curable conditions obscured by the mass of essential hypertensive patients. Based on early figures from tertiary referral centers of 6% for "surgically curable hypertension," a vigorous, costly, and sometimes dangerous workup was recommended and done on many patients, most of whom did not turn out to have a reversible disease. With time, it became apparent that the true prevalence of reversible hypertension in the community was probably less than 1% (76).

As has been pointed out by Sackett (77), these figures are different because of referral "filters" to which patients assessed at primary care level must pass before they are assessed in tertiary care centers. It is possible that we may now be seeing an analogous phenomenon with respect to the workup of dementia, especially in the elderly in whom the incidence and prevalence of Alzheimer's disease rises so rapidly with age (37, 71). Unless the incidence of reversible causes also rises as rapidly with age, the older the patient the less likely the dementia would be to reverse.

On the other hand, the opposite could be true. Some of the reversible causes could have been weeded out by the referring physicians, sending an artifically raised proportion of irreversible cases to the consultants. For example, in the study by Marsden and Harrison (7), all patients "had already been screened for an obvious cause for their dementia in the outpatient clinics of the hospital or at other hospitals" (emphasis mine). However, there are many reasons it is unlikely that many reversible causes would have been screened out before having reached the investigators whose work is surveyed here. First, reversible dementias often present atypically (21, 39, 43, 78), whereas Alzheimer's disease, the commonest cause of irreversible dementia, usually has a recognizable clinical course (79).

Second, in usual clinical practice, most cases of cognitive decline, if recognized, are referred for further workup. To further complicate the issue, there is evidence that even good physicians, both community (80) and hospital-based (81), frequently do not recognize which of their patients are demented. In addition, cases that are potentially reversible often are not recognized. For example, in the study by Fox and associates (9), the underlying diagnosis of potentially reversible disease in four of five cases referred to the investigators had not been previously suspected by the referring physicians.

Another source of selection bias is evident in the study (24) in which subjects were drawn exclusively from patients who had had CT scans. In another work (13), the study sample included not only patients admitted with a diagnosis of dementia, but also patients discharged whose case notes indicated a diagnosis known to cause dementia. This practice might have increased the proportion of potentially reversible cases reported.

Another potential bias was evident in most of the studies, in that no attempt was made to blind the investigators in determining "improvement." For example, in one case (9), a demented patient with a B12 deficiency "showed an initial rapid improvement in all of her dementia symptoms, but still had some evidence of organic brain syndrome at the time of discharge." These authors are to be commended for their candor, but it is hard to know exactly what "improvement" and "some evidence of organic brain syndrome" mean. How can this kind of findings be compared with findings reported in similar studies? Many authors followed this same practice. More appropriately, Larson and coworkers (28, 33) used an interesting consensus approach with predetermined criteria for reversibility.

Another possible bias concerns the sex ratio of study subjects. Table 1 shows that 47.9% of patients studied were female. This figure suggests that the researchers examined atypical samples, because as one adjusts for the different age-specific mortality of men and women (at least for Alzheimer's disease), either no sex differential (71) or a predominantly female one is reported (82). In absolute terms there are many more female than male patients with dementia; thus the relatively low percentage of women examined suggests a bias in the studies surveyed. (A partial explanation for this figure that does not eliminate the bias lies in the fact that six of the studies emanated from American Veterans' hospitals.)

A critical question is whether patients with "reversed" dementia stay well or relapse if followed up for long enough. For example, much has been said about depression being the cause of a reversible dementia (83). There are many well-documented cases of cognitive improvement after treatment of depression. For example, Rabins (39) reported on 16 patients who presented with co-existing dementia and major depression. Thirteen (81%) had a full recovery of cognition after treatment for their depression. Most maintained this improvement after 2 years of follow-up, but not all patients do so. In addition, some patients who at first seem to show improved cognition when depression is treated, ultimately manifest the presence of an underlying irreversible dementia that shows up only on follow-up. Reding and colleagues (84) report on 15 patients sent to a specialized dementia clinic initially judged to be depressed and given appropriate treatment for this condition. On follow-up, 8 (53%) developed progressive intellectual impairment.

To further complicate the matter, other studies (47, 85, 86) have indicated that the diagnoses of depression and dementia are not mutually exclusive. In a careful analysis of the validity of the concept of depressive pseudodementia, Patterson (47) suggests that there are good reasons to abandon the term, noting that all we can conclude is that dementing illnesses are frequently accompanied by rather than caused by depression. If the concept of depressive pseudodementia is not accepted, the number of potentially reversible causes of dementia would be reduced.

The phenomenon of "late occurrence" of cognitive deterioration is seen even in situations where early follow-up actually documented cognitive improvement.

Larson and associates (28) followed 13 patients with reversed dementia for more than 2 years (after treatment of "cause"). Surprisingly, even after the initial improvement, 62% (3 of 4 patients with hypothyroidism, 4 of 6 with medication toxicity, and 1 with subdural hematoma) developed progressive deterioration consistent with Alzheimer's dementia. Autopsy findings confirmed the presence of Alzheimer's disease in 2 patients.

Examining the problem from the point of view of the numerator only, Rabins (39) reported on 16 cases of potentially reversible dementia known to him. Of these, two thirds made a partial or full recovery. However, without knowing the denominator, it is difficult to draw any further conclusions from his data. Larson and associates (28) reported that of those cases initially labelled irreversible, 4.7% actually improved on follow-up. Others (38) report similar findings. Although these observations introduce a small note of optimism for patients given a diagnosis of irreversible dementia, there is no way to determine which of these cases actually will reverse.

Whatever the true incidence of reversible dementia turns out to be, one must remember that a zealous search for the individual with a rare cause can generate needless and not always harmless investigations. Thus, it may not be justifiable to subject large numbers of patients with irreversible dementia to a full diagnostic workup. There are many people with normal cognitive status (17) having needless testing as well. Receiving a false-positive diagnosis for reversible dementia can have serious consequences. For example, if a patient with dementia and his or her family are informed optimistically that the patient also has hypothroidism and that treatment will improve the state of cognition, one can imagine the distress engendered if the thyroid indices return to normal but the cognitive state does not.

More ominously, the diagnostic workup and attempts at treatment precipitated by a search for potentially reversible disease are not always benign. Several cases of serious morbidity secondary to the diagnostic workup as well as serious postoperative complications, including some deaths (subdural hematoma and normal pressure hydrocephalus), were reported (9, 10, 13). Perhaps not as dramatic, but equally disabling is the phenomenon of "labelling." For example, Garcia and coworkers (17) report the case of a woman with normal mental function who, after recovering from encephalitis, had a CT scan that indicated some atrophy. She was told that she would become demented; the "label" turned out to be disabling. Furthermore, it is well known that patients with aphasia or deafness can be mistakenly labelled as demented (18, 19), with presumably similar negative consequences. Knowing that the elderly are generally more fragile than their younger counterparts should encourage us to be as judicious as possible in our diagnostic recommendations, so as not to precipitate a harmful "cascade" effect (87) that will cause patients more harm than good.

With respect to dementia reversibility, the issue of alcohol abuse remains problematic. Although it is generally considered that abstinence will bring about some improvement in the Wernicke-Korsakoff syndrome, the clinical symptoms of Korsakoff disease do not usually abate (88). The treatment of chronic alcoholism is a difficult enough proposition even in younger people (88-90), with little work being reported on the elderly alcohol abuser. As the authors of one paper (24) in the series reviewed pointed out, it is difficult to separate social from excessive drinking and therefore to know exactly how to classify demented people with a history of alcoholism. It is for these reasons that dementias due to alcohol abuse were not included among the potentially reversible causes. However, it should be noted that Larson and coworkers (33) did describe three cases of partially reversed and one fully reversed dementia after treatment of alcohol abuse. If alcohol had been included in this category, the percentage of potentially reversible causes would have been slightly increased. (Because the cases of improvement with abstinence were included among cases with reversed disease, these figures have not changed.)

Table 4. Summary of 103 Cases Reported with Partially and Completely Reversed Dementia

Condition	Reversed Dementia		Total*	Cumulative %
	Partly	Completely		
	n	n	n (%)	n (%)
Drugs	17†	12	29(28.2)	28.2
Depression	18	9	27(26.2)	54.4
Metabolic	10	6	16(15.5)	69.9
Thyroid	6	1	7 (6.8)	...
B$_{12}$	...	1	1 (1.0)	...
Calcium	2	...	2 (1.9)	...
Hepatic	2	...	2 (1.9)	...
Other	...	4	4 (3.9)	...
Normal pressure hydrocephalus	8	3	11(10.7)	80.6
Subdural hematoma	5	1	6 (5.8)	86.4
Neoplasm	4	...	4 (4.0)	90.4
Other	9	1	10 (9.7)	100.0
Total‡	71(68.9)	32(31.1)	103 (100)	...

* Total includes all cases of partially and completely reversed dementia.
† Includes 4 cases of alcohol abuse.
‡ Figures are n(%).

As in the case of alcohol, the question remains as to whether multi-infarct dementia can be considered potentially reversible. Almost all authors surveyed here considered it to be an irreversible form of dementia. In none of 32 studies examined was any case of improvement of multi-farct dementia reported (Table 4). Nevertheless, many authors expressed the hope that it might be possible at least to stabilize, if not reverse, the disease by careful attention to stroke risk factors. In addition, there has been some recent evidence in support of the contention that multi-infarct dementia can be reversed. Meyer and colleagues (91) studied a cohort of 49 patients with multi-infarct dementia (mean age, 69.9 years), for an average of 22.2 months, with careful attention to blood pressure control. They found that among the 27 patients with hypertension (55%), patients who maintained their systolic blood pressure between 135 and 150 mm Hg showed either improvement or stabilization of their cognitive status.

Although these initial findings are promising, many factors argue against granting multi-infarct dementia full membership in the club of potentially reversible causes. First, not all patients with multi-infarct dementia have hypertension. Second, as reported by Meyer and colleagues (91), if the systolic blood pressure fell below 135 mm Hg, there were actual declines in both cognition and neurologic status. Blood pressure control in the elderly is difficult to achieve (92), and there is evidence that treatment in the very elderly may do more harm than good (93).

WORKUP AND TESTS

One must also consider what the economic cost of a workup would be for all patients. An extensive workup has been suggested by prestigious American consensus groups. For example, in 1980 a National Institute on Aging Task Force (42) recommended seven separate blood screens (including SMA 12), urinalysis, stool occult blood, chest roentgenogram, electrocardiogram, and CT scan. The National Institute on Aging group did not make a definitive recommendation for a full laboratory workup in all patients but suggested an "individualized approach." However, their statement leaves the reader with a fairly unambiguous message: "The relative ease of administering most of the tests required for the differential diagnosis and the gravity of dementia and delirium favor their early use" (emphasis mine).

More recently, a consensus conference sponsored by the National Institute on Aging and various other relevant National Institutes of Health, examined this question again (53). Although once again asserting that the careful history, physical and psychological examinations were primary, they listed a fairly extensive lab workup for all patients with "new onset" of dementia. Although similar to the 1980 National Institute on Aging recommendations (testing for stool occult blood was dropped), the CT scan was treated slightly differently. The wording of the recommendation left this reader unclear as to the group's exact intention. For this reason I quote it in its entirety: "Other ancillary tests that are appropriate in certain common situations are as follows: 1. Computed tomography of the brain (without contrast) is appropriate in the presence of history suggestive of a mass, or focal neurological signs, or in dementia of brief duration. Unless such diagnosis is obvious on first contact, computed tomography should be performed (53)."

It is my interpretation that this consensus group, although not recommending CT scans for all patients with new onset dementia, is suggesting that scans be done on most patients. Unfortunately, the two National Institutes of Health reports contained no references, and so it is difficult to know on what grounds (beyond the opinion of experts) these recommendations were made.

Cost must also be a consideration. Larson and associates (78) estimated that the 1980 National Institute on Aging (42) workup would cost approximately $668.50

51

per patient (1984 dollars) compared with his two more "selective test ordering strategies" of either $153.92 or $209.31 per patient. Larson's testing recommendations are based on the results of his own studies, an attempt to minimize costs, patient discomfort, and other morbidity due to testing, while making it extremely unlikely that one would miss a case of truly reversible dementia. It has been conservatively estimated that this approach could save hundreds of millions of dollars per year. The authors also point out that the recommended strategies must be effectively applied using sound clinical judgment. Otherwise, missed diagnoses and treatment opportunities would increase costs and human suffering that could outweigh any potential savings.

However, in a test of these selective strategies, Martin and coworkers (94, 95) found that "adoption of these reported strategies would have missed many diagnostic findings, primarily vascular insults and small meningiomas." Yet, it is unlikely that knowledge of these particular diagnoses would really have led to action that would have truly reversed the dementias in question.

An argument has been made (34, 96, 97) that even if reversible cases are not found, patients with dementia are still at high risk for other medical diseases, and these conditions should be sought after and treated vigorously. It is undoubtedly true that patients with dementia often present with serious coexisting pathology, but this finding does not necessarily imply that each patient with dementia needs a full and costly lab workup to look for occult diseases not shown to reverse dementia. Rather, these patients should be considered at high risk with respect to their general health and offered appropriate therapy. The goal should be to make the patient more comfortable or functional rather than to try (usually in vain) to reverse dementia.

It must always be remembered that investigators of the causes of and therapy for the dementias do require "early" cases for their work. In this particular case and for other research purposes, a fairly exhaustive workup is required to assure that the diagnosis is correct. However, a critical review of the literature does not support a translation of this research requirement into the need for clinicians at the primary and secondary level to subject all demented patients to such an exhaustive workup.

CONCLUSION

A critical review of studies examining the causes for dementia was done. Although dementia is primarily a problem of the elderly, many of the papers reviewed neither adequately reported the age structure of the study sample, nor included many older subjects. In addition, many studies reported only potentially reversible causes, rather than indicating whether on follow-up cognitive function actually improved with treatment. In 11 studies that did do appropriate follow-up and documented improvement or lack thereof, a lower figure for fully reversible disease (3%) than is usually reported in the literature was calculated. Many of the studies that did report proper follow-up had other biases; the main bias was "referral filter."

In those few cases where reversibility was initially documented, many of these patients showed renewed cognitive decline on follow-up, indicating the presence of an underlying irreversible dementia. In grappling with this problem, Larson and associates (28) suggest that the concept of reversible dementia may have limited practical significance and that one should not concentrate on the notion of a reversible-irreversible treatable-untreatable dichotomy.

Although some promising work has recently been reported (98) with respect to the treatment of the most common irreversible cause of dementia, Alzheimer's disease, the improvement reported was palliative, not curative (99), and some criticisms of the research methodology have been made (100). It remains to be seen if these hopeful reports will be confirmed elsewhere. **Even if a new and efficacious treatment of Alzheimer's disease, analogous to that existing for Parkinson's disease, is seen to work, an approach to all new cases of dementia that seeks out reversible causes will still be needed. For most reversible causes, the workup can be appropriately and economically done starting with a searching history and guided physical examination, followed by a selective and hierarchical choice of lab tests.

In the excitement of the chase, it must never be forgotten that most patients, who will in the end be shown to have an irreversible cause for their dementia, still require support and care (101). As has been pointed out elsewhere (43, 102), the physician must remember that his job is always to "care" for the demented patient; the opportunity for a true cure seldom arises.

Although not always simple to effect in older subjects (72, 103), further research on the subject is still needed. Hopefully it will be designed with an eye toward avoiding some of the biases and errors pointed out above. Cases should be generated from community-based rather than institutional samples. Standardized terminology and testing protocols as well as appropriate follow-up will help make studies comparable and capable of answering the question of what is the true incidence of reversible dementia. As has been so elegantly expressed by Mulley (44): "To deny demented patients comprehensive assessment is neglect, to subject them all to detailed investigation is unnecessary."

*Reproduced, with permission, from: A.M. Clarfield, MD, The Reversible Dementias: Do They Reverse? *Ann. Intern. Med.* 1988;109:476-486.

**Author's note: They have been recently refuted; see Gauthier, S. Bouchard, R., Lamontagne, A., et al. Tetrahydroaminoacridine-Lecithin combination treatment in intermediate stage Alzheimer's disease: Results of a Canadian double-blind crossover multicentre trail (1990). *In English Ed., 322*: 1272-76.

INTRODUCTION REFERENCES

1. National Institute on Aging Task Force: Senility reconsidered: treatment possibilities for mental impairment in the elderly (1980). *JAMA, 244*:259-263.
2. Council on Scientific Affairs, American Medical Association: Dementia (1986). *JAMA, 256*:2234-2238.
3. NIH Consensus Conference: Differential diagnosis of dementing diseases (1987). *JAMA, 258*:3411-3416.
4. Royal College of Physicians Committee on Geriatrics. Organic mental impairment in the elderly: implications for research, education, and the provision of services (1981). J. R. College of Physicians, London, *15*:141-67.
5. Larson, E. G., Reifler, B. V., Sumi, S. M. (1986). Diagnostic tests in the evaluation of dementia: a prospective study of 200 elderly outpatients. *Arch. Intern. Med., 126*:1917-1922.
6. Clarfield, A. M. (1988). The reversible dementias: do they reverse? *Ann. Intern. Med., 109*:476-486.
7. Clarfield, A. M. (1989). Diagnostic assessment of dementia. [Letter]. *Ann. Intern. Med., 110*:670.
8. Wilson, D. B., Guyatt, G. H. Streiner, D. L. (1987). The diagnosis of dementia. *Can. Med. Assoc. J., 137*:625-629.
9. Barry, P. P., Moskowitz, M. A. (1988). The diagnosis of reversible dementia in the elderly: a critical review. *Arch. Int. Med., 148*: 1914-1918.
10. Sackett, D. L. (1979). Biases in analytic research. *J. Chronic Dis., 32*:51-63.
11. Clarfield, A. M., Friedman, R. (1985). A survey of the age structure of "age-relevant" articles in four medical journals. *J. Amer. Ger. Soc., 33*:773-778.

12. Evans, D. A., Funkenstein, H. H., Albert, M. S. (1989). Prevalence of Alzheimer's disease in a community population of older persons. *JAMA, 262*:2551-56.

13. Canadian Medical Association (1987). The elderly: Challenges for today - options for the future. Report of the Committee on Health Care of the Elderly. Canadian Medical Association, Ottawa.

14. Clarfield, A. M., Larson, E. B. (1990). Should a major imaging procedure (CT or MRI) be required in the workup of dementia? An opposing view. *J. Family Practice, 31*.

15. Katzman, R. (1990). Should a major imaging procedure (CT or MRI) be required in the workup of dementia? An affirmative view. *J. Family Practice, 31*.

16. Rhymes, J. A., Woodson, C., Sparage-Sachs, R., Cassel, C. (1989). Non-medical complications of diagnostic workup for dementia. *J. Amer. Ger. Soc., 37*:1157-64.

REFERENCES

1. Transactions of the Clinical Society of London. (1988). Supplement to Volume 21. Report of a Committee of the Clinical Society of London to investigate the subject of myxoedema. London: Longman Green.

2. Hahn, R. D., Webster, B., Weickhardt, G. (1959). Penicillin treatment of general paresis (dementia paralytica). *Arch. Neurol., 91*:557-90.

3. Adams, R. D., Fisher, C. M., Hakim, S. (1965). Symptomatic occult hydrocephalus with "normal" cerebrospinal fluid pressure. *New Engl. J. Med., 273*:117-26.

4. Haase, G. R. (1977). Diseases presenting as, dementia. In: Wells, C. E. (ed) *Dementia,* 2nd ed. Philadelphia: F. A. Davis, 26-67.

5. Hutton, J. T. (1981). Atrial myxoma as a cause of progressive dementia. *Arch. Neurol., 38*:533.

6. Roos, R. P., Johnson, R. T. (1977). Viruses and dementia. *Contemp. Neurol. Ser., 15*:93-112.

7. Marsden, C. D., Harrison, M. J. (1972). Outcome of investigation of patients with presenile dementia. *Br. Med. J., 2*:249-52.

8. Pearce, J., Miller, E. (1973). *Clinical Aspects of Dementia.* London: Balliere Tindall, 81.

9. Fox, J. H., Topel, J. L., Huckman, M. S. (1975). Dementia in the elderly-- a search for treatable illnesses. *J. Gerontol., 30*:557-64.

10. Freemon, F. R. (1976). Evaluation of patients with progressive intellectual deterioration. *Arch. Neurol., 33*:658-9.

11. Harrison, J. J., Marsden, C. D. (1977). Progressive intellectual deterioration [Letter]. *Arch. Neurol., 34*:199.

12. Katzman, R (1975). Personal Communication. In: Wells, C. E., (ed.) *Dementia,* 2nd ed. Philadelphia: Davis, F. A., 250.

13. Victoratos, G. C., Lenman, J. A., Herzberg, L. (1977). Neurological investigation of dementia. *Br. J. Psychiatry, 130*:131-3.

14. Kokmen, E., Okazaki, H., Schoenberg, B. S. (1980). Epidemiologic patterns and clinical features of dementia in a defined U.S. population. Trans. *Am. Neurol. Assoc., 105*:334-6.

15. Reifler, B. V., Eisdorfer, C. (1980). A clinic for impaired elderly and their families. *Am. J. Psychiatry, 137*:1399-403.

16. Barnes, R. F., Raskind, M. A. (1981). DSM-III criteria and the clinical diagnosis of dementia: a nursing home study. *J. Gerontol., 36*:20-7.

17. Garcia, C. A., Reding, M. J., Blass, J. P. (1981). Overdiagnosis of dementia. *J. Am. Geriatr. Soc., 29*:407-10.

18. Hutton, J. T. (1981). Results of clinical assessment for the dementia syndrome: Implications for epidemiologic studies. In: Schuman, L. M., Mortimer, J. A. (eds). *The Epidemiology of Dementia.* New York: Oxford University Press: 62-9.

19. Hutton, J. T. (1981). Senility reconsidered [Letter]. JAMA, 245:1025-6.

20. Rabins, P. V. The prevalence of reversible dementia in a psychiatric hospital. *Hosp. Community Psychiatry, 32*:490-2.

21. Smith, J. S., Kiloh, L. G. (1981). The investigation of dementia: results in 200 consecutive admissions. *Lancet, 1*:824-7.

22. Benson, D. F., Cummings, J. L., Tsai, S. Y. (1982). Angular gyrus syndrome simulating Alzheimer's disease. *Arch. Neurol., 39*:616-20.

23. Delaney, P. (1982). Dementia: The search for reversible causes. *South Med. J., 75*:707-9.

24. Freemon, F. R., Rudd, S. M. (1982). Clinical features that predict potentially reversible progressive intellectual deterioration. *J. Am. Geriatr. Soc., 30*:449-51.

25. Maletta, G. J., Pirozzolo, F. J., Thompson, G., Mortimer, J. A. (1982). Organic mental disorders in a geriatric outpatient population. *Am. J. Psychiatry, 139*:521-23.

26. Sabin, T. D., Vitug, A. J., Mark, V. H. (1982). Are nursing home diagnosis and treatment adequate? *JAMA, 248*:321-2.

27. Martin, B. A., Thompson, E. G., Eastwood, M. R. (1983). The clinical investigation of dementia. *Can. J. Psychiatry, 28*:282-6.

28. Larson, E. B., Reifler, B. V., Featherstone, H. J., English, D. R. (1984). Dementia in elderly outpatients: A prospective study. *Ann. Intern. Med., 100*:417-23.

29. Roca, R. P., Klein, L., McArthur, J. C. (1984). Treatable conditions among demented medical inpatients [Abstract]. *Clin. Res.,* 300A.

30. Folstein, M., Anthony, J. C., Parhad, I., Duffy, B., Gruenberg, E. M. (1985). The meaning of cognitive impairment in the elderly. *J. Am. Geriatr. Soc., 33*:228-35.

31. Gilchrist, P. N., Rozenbilds, U. Y., Martin, E., Connolly, H. (1985). A study of 100 consecutive admissions to a psychogeriatric unit. *Med. J. Aust., 143*:236-7.

32. Hammerstrom, D. C., Zimmer, B. (1985). The role of lumbar puncture in the evaluation of dementia: the University of Pittsburgh study. *J. Am. Geriatr. Soc., 33*:397-400.

33. Larson, E. B., Reifler, B. V., Sumi, S. M., Canfield, C. G., Chinn, N. M. (1985). Diagnostic evaluation of 200 elderly outpatients with suspected dementia. *J. Gerontol., 40*:536-43.

34. Renvoize, E. G., Gaskell, R. K., Klar, H. M. (1985). Results of investigation in 150 demented patients consecutively admitted to a psychiatric hospital. *Br. J. Psychiatry., 147*:204-5.

35. Schoenberg, B. S., Anderson, D. W., Haerer, A. F. (1985). Severe dementia: prevalence and clinical features in a biracial US population. *Arch. Neurol., 42*:740-3.

36. Erkinjuntti, T., Wikstrom, J., Palo, J. Autio, L. (1986). Dementia among medical inpatients: Evaluation of 2000 consecutive admissions. *Arch. Intern. Med., 146*:1923-6.

37. Sayetta, R. B. (1986). Rates of senile dementia, Alzheimer's type, in the Baltimore Longitudinal Study. *J. Chronic Dis., 39*:271-86.

38. Pfeffer, R. I., Afifi, A. A., Chance, J. M. (1987). Prevalence of Alzheimer's disease in a retirement community. *Am. J. Epidemiol., 125*:420-36.

39. Rabins, P. V. (1985). The reversible dementias. In: Arie, T. (ed). *Recent Advances in Psychogeriatrics*. Edinburgh: Churchill Livingstone. 93-102.

40. Wlodarczyk, D. M. (1985). Dementia: guidelines for improving Tx. *Geriatrics., 40*:35-45.

41. Hoffman, R. S. (1982). Diagnostic errors in the evaluation of behavioral disorders. *JAMA, 248*:964-7.

42. National Institute on Aging Task Force (1980). Senility reconsidered: treatment possibilities for mental impairment in the elderly. *JAMA, 244*:259-63.

43. Arie, T. (1973). Dementia in the elderly: diagnosis and assessment. *Br. Med. J., 4*:540-3.

44. Mulley, G. P. (1986). Differential diagnosis of dementia [Editorial]. *Br. Med. J. [Clin. Res.], 292*:1416-8.

45. Royal College of Physicians Committee on Geriatrics (1981). Organic mental impairment in the elderly: implications for research, education and the provision of services. *J. R. Coll. Physicians Lond., 15*:141-67.

46. McIntyre, L., Frank, J. (1987). Evaluation of the demented patient. *J. Family Pract., 24*:399-404.

47. Patterson, C. (1986). The diagnosis and differential diagnosis of dementia and pseudo-dementia in the elderly. *Can. Fam. Physician, 32*:2607-10.

48. Lister J. Shattuck lecture--The politics of medicine in Britain and the United States. *New Engl. J. Med., 315*:168-74.

49. Evans, J. G. (1982). Anglo-American differences in care for the elderly: reflections on a visiting professorship. *J. Am. Geriatr. Soc., 30*:348-51.

50. Wells, C. E. (1978). Chronic brain disease: An overview. *Am. J. Psychiatry, 135*:1-12.

51. McKhann, G., Drachman, D., Folstein, M., Katzman, R., Price, D., Stadlan, E. M. (1984). Clinical diagnosis of Alzheimer's disease: Report of NINCDS-ADRDA Work Group under the auspices of Department of Health and Human Services Task Force on Alzheimer's Disease. *Neurology, 34*:939-44.

52. Blass, J. P., Barclay, L. I. (1985). New developments in the diagnosis of the dementias. *Drug Develop. Res., 5*:39-58.

53. Consensus conference: differential diagnosis of dementing diseases (1987). *JAMA, 258*:3411-6.

54. Dementia in Later Life: Research and Action. (1986). Report of a WHO Scientific Group on Senile Dementia. No. 730. Geneva: World Health Organization.

55. Cowell, D. D. (1983). Senile dementia of the Alzheimer's type: A costly problem [Editorial]. *J. Am. Geriatr. Soc., 31*:61.

56. Beck, J. C., Benson, D. F., Scheibel, A. B., Spar, J. E., Rubenstein, L. Z. (1982). Dementia in the elderly: The silent epidemic. *Ann. Intern. Med., 97*:231-41.

57. Sourander, P., Sjogren, H. (1970). The concept of Alzheimer's disease and its clinical implications. In: Wolstenholme, G. E. W., O'Connor, M. (eds). *Alzheimer's and Related Conditions.* London: Churchill, 11-36.

58. Tomlinson, B. E., Blessed, G., Roth, M. (1970). Observations on the brains of demented old people. *J. Neurol. Sci., 11*:205-42.

59. Malamud, N. (1972). Neuropathology of organic brain syndromes associated with aging. In: Gaitz, C. M. (ed). *Aging and the Brain.* New York: Plenum, 63-87.

60. Todorov, A. B., Go, R. C., Constantinidis, J. Elston, R. C. (1975). Specificity of the clinical diagnosis of dementia. *J. Neurol. Sci., 26*:81-98.61.

61. Ojeda, V. J., Mastaglia, F. L., Kakulas, B. A. (1986). Causes of organic dementia: a necropsy survey of 60 cases. *Med. J. Aust., 145*:69-71.

62. Kokmen, E., Offord, K. P., Okazaki, H. (1987). A clinical and autopsy study of dementia in Olmsted County, Minnesota, 1980-81. *Neurology, 37*:426-30.

63. Shaw, G. M., Harper, M. E., Hahn, B. H. (1985). HTLV-III infection in brains of children and adults with AIDS encephalopathy. *Science, 227*:177-82.

64. *Diagnostic and Statistical Manual of Mental Disorders III.* (1980). Washington, D.C.: American Psychiatric Association.

65. Nott, P. N., Fleminger, J. J. (1975). Presenile dementia: the difficulties of early diagnosis. *Acta Psychiatr. Scand., 51*:210-7.

66. Ron, M. A., Toone, B. K., Garralda, M. E., Lishman, W. A. (1979). Diagnostic accuracy in presenile dementia. *Br. J. Psychiatry, 134*:161-8.

67. Wilson, D. B., Guyatt, G. H., Streiner, D. L. (1987). The diagnosis of dementia. *Can. Med. Assoc. J., 137*:625-9.

68. Cummings, J., Benson, F., Loverme, S., Jr. (1980). Reversible dementia: illustrative cases, definition, and review. *JAMA, 243*:2434-9.

69. Duckworth, G. S., Ross, H. (1975). Diagnostic differences in psychogeriatric patients in Toronto, New York and London, England. *Can. Med. Assoc. J., 112*:847-51.

70. Smith, J. S., Kiloh, L. G., Ratnavale, G. S., Grant, D. A. (1976). The investigation of dementia: the results in 100 consecutive admissions. *Med. J. Aust., 2*:403-5.

71. Mortimer, J. A., Schuman, L. M., French, L. R. (1981). Epidemiology of dementing illness. In: Schuman, L. M., Mortimer, J. . *The Epidemiology of Dementia*, New York: Oxford University Press, 3-23.

72. Clarfield, A. M., Friedman, R. (1985). Survey of the age structure of "age-relevant" articles in four general medical journals. *J. Am. Geriatr. Soc., 33*:773-8.

73. Sackett, D. I., Haynes, R. B., Tugwell, P. (1985). *Clinical Epidemiology: A Basic Science for Clinical Medicine.* Boston: Little Brown and Co., 289-90.

74. Cox, S. (1983). Comments on "unsolved issues" and reversible dementia [Letter]. *J. Am. Geriatr. Soc., 31*:126.

75. Wolff, M. L. (1982). Reversible intellectual impairment: an internist's perspective. *J. Am. Geriatr. Soc., 30*:647-50.

76. Rudnick, K. V., Sackett, D. L., Hirst, S., Holmes, C. (1977). Hypertension in a family practice. *Can. Med. Assoc., 117*:492-7.

77. Sackett, D. L. (1979). Biases in analytic research. *J. Chronic Dis., 32*:51-63.

78. Larson, E. B., Reifler, B. V., Sumi, S. M., Camfield, C. G., Chinn, N. M. (1986). Diagnostic tests in the evaluation of dementia: a prospective study of 200 elderly outpatients. *Arch. Intern. Med., 146*:1917-22.

79. Reisberg, B. (1983). Clinical diagnosis and differential diagnosis of Alzheimer's disease and related disorders. In: Reisberg, B. (ed). *Alzheimer's Disease.* New York: The Free Press, 173-87.

80. Williamson, J., Stokoe, J. H., Gray, S. (1964). Old people at home: their unreported needs. *Lancet, 1*:117-20.

81. McCartney, J. R., Palmateer, L. M. (1985). Assessment of cognitive deficit in geriatric patients: a study of physical behavior. *J. Am. Geriatr. Soc., 33*:467-71.

82. Schoenberg, B. S. (1986). Epidemiology of Alzheimer's disease and other dementing illnesses. *J. Chronic Dis., 39*:1095-104.

83. Wells, C. E. (1979). Pseudodementia. *Am. J. Psychiatry, 136*:895-900.

84. Reding, M., Haycox, J., Wigforss, K., Brush, D., Blass, J. P. (1984). Follow-up of patients referred to a dementia service. *J. Amer. Geriatr. Soc., 32*:265-8.

85. Reifler, B. V. (1982). Arguments for abandoning the term pseudodementia. *J. Am. Geriatr. Soc., 30*:665-8.

86. Reifler, B., Larson, E., Hanley, R. (1982). Coexistence of cognitive impairment and depression in geriatric outpatients. *Am. J. Psychiatry, 139*:623-6.

87. Mold, J. W., Stein, H. F. (1986). The cascade effect in the clinical care of patients. *New Engl. J. Med., 314*:512-4.

88. West, L. J., Maxwell, D. S., Noble, E. P., Solomon, D. H. (1984). Alcoholism. *Ann. Intern. Med., 100*:405-16.

89. Blume, S. B. (1983). Is alcoholism treatment worthwhile? *Bull. N.Y. Acad. Med., 59*:171-80.

90. Vaillant, G. E., Clark, W., Cyrus, C. (1983). Prospective study of alcoholism treatment: Eight year follow-up. *Am. J. Med., 75*:455-63.

91. Meyer, J. S., Judd, B. W., Tawakina, T., Rogers, R. L., Mortel, K. F. (1986). Improved cognition after control of risk factors for multi-infarct dementia. *JAMA, 256*:2203-9.

92. Larochelle, P. (ed) (1985). *Report of the Consensus Development Conference in the Management of Hypertension in the Elderly in Canada. Montreal: Canadian Hypertension Society.*

93. Amery, A., Birkenhager, W., Brixko, R. (1986). Efficacy of antihypertensive drug treatment according to age, sex, blood pressure and previous cardiovascular disease in patients over the age of 60. *Lancet, 2*:589-92.

94. Martin, D. C., Miller, J., Kapoor, M., Karpf, M. (1985). A test of diagnostic strategies in senile dementia. [Abstract]. *Clin. Res., 33*:726A.

95. Martin, D. C., Miller, J., Kapoor, W., Karpf, M., Boller, F. (1987). Clinical prediction rules for computed tomographic scanning in senile dementia. *Arch. Int. Med., 147*:77-80.

96. Larson, E. G., Buchner, D. M., Uhlmann, R. F., Reifler, B. V. (1986). Caring for elderly patients with dementia [Editorial]. *Arch. Intern. Med., 146*:1909-10.

97. Uncovering physical illness in elderly patients with dementia [Editorial] (1977). *Br. Med. J., 2*:1499-1500.

98. Summers, W. K., Majovski, L. V., Marsh, G. M., Tachiki, K., Kling, A. (1986). Oral tetrahydroaminoacridine in long-term treatment of senile dementia, Alzheimer's type. *New Engl. J. Med., 315*:1241-5.

99. Davis, K. L., Mohs, R. C. (1986). Cholinergic drugs in Alzheimer's disease [Editorial]. *New Engl. J. Med., 315*:1286-7.

100. Pirozzolo, F. J., Hermann, N., Small, G. W., Tariot, P. N., (1987). Oral tetrahydroaminoacridine in the treatment of senile dementia, Alzheimer's type [Letters]. *New Engl.J. Med., 316*:1603-5.

101. Arie, T. (1986). Management of dementia: A review. *Br. Med. Bull., 42*:91-6.

102. Cassel, C. K., Jameton, A. L. (1981). Dementia and the elderly: an analysis of medical responsibility. *Ann. Intern. Med., 94*:802-7.

103. Zimmer, A. W., Calkins, E., Hadley, E., Ostfeld, A. M., Kaye, J. M. Kaye, D. (1985). Conducting clinical research in geriatric populations. *Ann. Intern. Med., 276*-83.

THE DIAGNOSIS OF ALZHEIMER'S DISEASE

Robert A. Murden, M.D.

Assistant Professor, Department of Medicine
University of Kansas Medical Center
Kansas City, Kansas

The diagnosis of Alzheimer's Disease has traditionally been a diagnosis of exclusion. Although there is some modestly promising research on diagnostic testing in Alzheimer's, this remains the standard currently, although clinical patterning can be useful additional evidence. This discussion will focus on this traditional diagnosis of exclusion in the following four-step approach. The first step is to ensure that the patient does, indeed, have dementia. The second step is to review the differential diagnosis of dementia and examine for all of the causes other than Alzheimer's. The third step is to decide if the clinical presentation of the patient under consideration is consistent with one of the several clinical patterns of Alzheimer's. The fourth step is to avoid certain pitfalls of diagnosis. These steps shall be discussed in detail in the order noted above.

DIAGNOSING DEMENTIA

The initial step is to determine that the patient truly has dementia as distinguished from delirium, psychiatric illness, or benign senescent memory loss. In establishing standardized definitions for Alzheimer's Disease for research purposes, the NINCDS-ADRDA (National Institute for Neurological and Communicative Disorders and Stroke-Alzheimer's Disease and Related Disorders Association) included what was basically a four criteria definition for dementia. (1) These criteria are: 1) a history of a decline in cognitive functioning; 2) an abnormal score on a standardized neuropsychological test; 3) neuropsychiatric abnormalities in at least two different cognitive areas; and 4) the absence of delirium. This last criterion, excluding patients who are delirious with stupor or clouded consciousness, is relatively obvious and requires no elaboration. This definition of dementia is essentially equivalent to the DSM-III diagnosis of dementia, with the exception that DSM-III requires an etiologic factor to be determined.

The first criterion, the history of decline in cognitive function, is perhaps the most important. A great deal of time should be spent in obtaining this history. A person other than the patient should be interviewed, and this should be preferably someone who has known the patient for a fair amount of time. Many detailed questions should be asked as to which cognitive deficits are present, when each started, and how each progressed. For example, if the patient can no longer cook for herself, why can't she, when and why did this inability start, and has there been any progression? If only a few losses are volunteered, other functions such as cooking,

New Directions in Understanding Dementia and Alzheimer's Disease,
Edited by T. Zandi and R. J. Ham, Plenum Press, New York, 1990

59

getting to the store and back, handling finances, remembering relatives' names, etc., should be inquired about. To fulfill this first criterion for dementia, the patient must exhibit significant cognitive losses from a previously well baseline.

The second criterion is an abnormal score on a standardized neuropsychological test. There are several such tests available, the most widely used of which is the Mini-Mental State Exam of Folstein (2) (see Chapter 1, Illustration 1). This is a 30 point test in which a score of 23 or less traditionally suggests dementia. In the Brooklyn, New York Alzheimer's Disease Assistance Center we administer this test along with two other brief exams in a total of about 15 minutes. The first of the other two tests is carried out by having the patient draw freehand a picture of a clock, putting in all the numbers and placing the hands so that the time reads 3 o'clock. The last test is to ask the patient to name all of the words (excluding proper names) he/she can think of which start with the letter S. The patient is given 60 seconds and 12 words or better is considered normal. Poorly educated individuals are asked to name all of the items which might be found in a refrigerator with the same timing and scoring requirement.

Patients who score 23 or less on the Mini-Mental State fulfill the second criterion for dementia. If the patient also performs poorly drawing the clock and generating the list of words, that is further confirmation of an abnormal mental status.

The third criterion for dementia is that the mental status changes are in at least two different cognitive areas. The cognitive areas include orientation, memory, language, copying, attention, calculation, abstraction, naming, and word generation. The above mentioned mental status testing (Mini Mental State, clock drawing, word list generation) examines all of these areas except abstraction; therefore, poor performance in at least two of these areas would fulfill the third criterion. An isolated loss of language or memory, for instance, may be a very early sign of dementia, but is more likely an isolated lesion of some sort and not a true global dementia.

Patients who fulfill all four criteria are regarded as having dementia and the next step, exploring the differential diagnosis of dementia, is begun. Patients who fulfill no criteria are clearly not demented. Patients who fulfill some, but not all criteria, cannot be regarded as definitely demented and may need to be followed clinically for a while before a diagnosis can be achieved with confidence.

EXAMINE DIFFERENTIAL DIAGNOSIS OF DEMENTIA

Once the diagnosis of dementia is established, the differential needs to be reviewed one by one and testing or clinical criteria applied for each possibility until it is accepted or rejected. This will be approached in a step-wise manner. Some of the causes of dementia are labelled irreversible, and some are thought to be reversible, treatable, or arrestable. Much of the importance of this second step of examining the differential diagnosis lies with the possibility of uncovering a treatable, reversible or arrestable cause. The irreversible causes will be discusssed first, followed by the much more rare, potentially reversible causes.

A mnemonic will help in remembering the various causes of dementia. A useful one which breaks down the causes is the following:

> MAMA'S
> BITTEN
> DAD

with MAMA's reflecting the irreversible causes, DAD reflecting the most important reversible causes, and BITTEN including the most rare, potentially reversible causes.

MAMA's stands for Multi-infarct dementia, Alzheimer's, Mixed, Alcoholic dementia, and Subcortical dementias. Multi-infarct dementia is primarily diagnosed clinically. A history of a stepwise decline in cognitive losses (rapid decline over a short period, plateau, another rapid decline, more plateau, etc.) is the hallmark of this diagnosis. A labile personality is often seen, and this disorder may be seen in younger patients (50s or 60s) who are predisposed to strokes with a history of hypertension or atherosclerosis. A clear history of strokes at the times of decline and a CT showing some type of infarcts is helpful additional data but is not mandatory.

Alzheimer's disease will be discussed later on in this chapter. Mixed dementia refers to people who have features of both multi-infarct dementia and Alzheimer's. Alcoholic dementia is also a diagnosis of exclusion of sorts. If a person has a history of heavy alcohol use, all causes other than Alzheimer's have been ruled out, and there is significant cortical atrophy on CT scan, the diagnosis of alcoholic dementia can be made.

Subcortical dementias are a group of diseases including Parkinson's disease with dementia, Huntington's chorea, Wilson's disease, and Binswanger's disease. These dementias are characterized by extreme slowness of speech and thought processes (taking two minutes to answer "What month is this?", then answering correctly, is classical) and by some sort of movement or gait disturbance. Binswanger's disease also has characteristic CT abnormalities.

The next group of possibilities includes the uncommon potentially reversible causes of dementia. The first is B12 deficiency, or pernicious anemia, which rarely causes dementia and can be ruled out with a normal serum B-12 level. The second cause is Infections, which includes the chronic meningidites of, cryptococcus, and tuberculosis, as well as neurosyphilis. A prior study has shown (3) that people with chronic meningitis as the cause for their dementia will have a history of fever and a rapid course (3 months or less) for their mental status changes. Therefore, those with fever and/or a rapid course should have a lumbar puncture as a part of the evaluation. This study also suggested that only one out of every 200 people with a positive peripheral serology (VDRL or FTA) and dementia will have neurosyphilis as the cause, and these will always have a rapid course. (4) Thus again, only those with a rapid onset for their dementia will require a lumbar puncture for evaluation.

The Ts stand for Trauma and Tumor and a CT scan is useful in the evaluation for dementia primarily to search for evidence of chronic subdural hematomas or slow growing tumors such as meningiomas. A CT scan is also useful in searching for Normal pressure hydrocephalus as a cause for the dementia. A negative CT rules out these possible etiologies completely. Finally, Endocrine causes, primarily hypothyroidism, should be considered and thyroid function tests should be ordered.

The final differential diagnostic categories are the more common reversible causes of dementia: Drugs, All other causes, and Depression. Many medications can cause altered mental status, including hypnotics, tranquilizers, anti-psychotics, pain medicines, cimetidine, toxic levels of anti-convulsants or anti-asthmatics, excess hypoglycemics, etc. Reducing or eliminating these drugs whenever possible is to be encouraged.

Depression can either mimic dementia or be a concomitant of dementia. When a demented person appears depressed, it is useful to treat the person for depression. This may result in no, partial, or total improvement.

This one-by-one examination of the differential diagnosis of dementia can have two results. The first is that one of these causes can be strongly suggested by clinical and laboratory findings, and the diagnosis is made. The second is that none of these causes will be suggested by the review, and therefore the diagnosis of probable Alzheimer's disease is made by exclusion. If this is the case, the next step should be taken in order to have more confidence with this diagnosis.

CLINICAL PATTERNS SUGGESTING ALZHEIMER'S

Once a diagnosis of probable Alzheimer's disease is made by excluding or ruling out all other possible causes, the following clinical patterns should be looked for in order to add confidence to the diagnosis. These patterns are found in the clinical progression of the disease, the sequence of progression of mental status abnormalities, and the patterns of personality changes.

The clinical progression of Alzheimer's disease involves three basic patterns. The classic pattern, accounting for about half of all Alzheimer's victims (or 1/4 of all patients with dementia), is a slowly progressive decline in function over several months to a few years. In this form the family will always state that the person is worse off than he/she was 6-12 months earlier. This pattern typically follows three stages (early, middle, late) where the individual progresses from a fluctuating mental status with some understanding to a severely helpless state with troubles eating or even walking.

The second pattern is that of a prolonged plateau phase. In this case there is a slow-to-rapid decline sometime in the past, but the person is currently functioning the same as 6-12 months ago. This plateau can last from 2-4 years before deterioration in functioning begins again. It is often difficult to distinguish plateau phase Alzheimer's from multi-infarct dementia not having repeat infarcts or from alcoholic dementia with cessation of drink.

The third pattern is that of a very rapid and severe decline in functioning. This is the least common pattern (seen in perhaps 15% with Alzheimer's) and is seen most often in the younger Alzheimer's victim. This is a person under age 65, with no alcohol history, and with a rapid decline to severe dementia in 1-2 years. The presence of one of these three patterns in a patient previously thought to have Alzheimer's disease, when other possible causes of dementia were ruled out, gives further confidence to the diagnosis.

The second issue is that of the sequence of mental status losses. The typical Alzheimer's patient initially loses memory and orientation, with word generation and copying abilities failing relatively early, and attention and language being preserved until late in the course of the disease. The presence of such a pattern of losses again adds confidence to the diagnosis of probable Alzheimer's, whereas the finding of a person with early language and attention deficits but fairly well preserved orientation would suggest a different diagnosis.

Finally, certain personality changes are likely to be seen in Alzheimer's disease. Most commonly the patient has a general loss of personality characteristics, becoming a hollow shell of sorts. A relatively unchanged personality would be less common but consistent with Alzheimer's, as would major personality changes with paranoid tendencies. On the other hand, a very labile affect and labile personality would be most consistent with multi-infarct dementia, and a very stable personality with sudden severe paranoia would be more consistent with drugs, depression or tumors.

PITFALLS IN THE DIAGNOSIS

One common problem occurs when assessing mental status changes in poorly educated individuals. Although two parts of our four part diagnosis of dementia are unaffected (absence of delirium and history of decline in cognitive function), the scoring on mental status testing can be affected by education and people can score in the dementia range when they are, in fact, non-demented.

Previous studies have shown Mini-Mental State Exam scores to be affected by education. (2,5,6) These large population studies have shown higher percentages of people with an 8th grade education or less score 23 or below when compared with those with more education. An ongoing longitudinal study of this issue is being carried out at the Brooklyn Alzheimer's Disease Assistance Center. In this study, people who initially score 23 or less on the Mini-Mental State but who are clinically non-demented, are followed with repeat Mini-Mental State exams and clinical exams. About 350 people are involved in this study to date.

To summarize our findings, non-demented individuals with an 8th grade or less education score an average of 25.4 out of 30 on the Mini-Mental State, versus a mean score of 28.2 out of 30 in those with a 9th grade or better education. In the well educated without dementia, less than 2% will score 23 or below. In contrast, 1/4 of the poorly educated without dementia may score 23 or less, all in the 18-23 range. There was no difference between blacks and whites in total scores, individual item scores, or percent scoring 23 or less, when education was kept constant. On the other hand, many individual items of the Mini-Mental State (including orientation, attention, recall, language, and copying items) were significantly correlated with education. Of the 50 individuals scoring 23 or less but not clinically demented at initial presentation, 2 of 3 who were well educated were found to be demented on follow-up and the third did not return for repeat testing. Of the 48 with low education levels, 10 were lost to follow-up and 8 of the remaining 38 (21%) were classified as demented after longitudinal assessment.

Our recommendations based on this study and the previous work in this area is that in people with an 8th grade education or less, a Mini-Mental State score of 17 or less always defines dementia, but that a score of 18-23 should be considered equivocal when using the four-part diagnosis of dementia. In such individuals we recommend longitudinal follow-up with a diagnosis of dementia being applied if clear functional and Mini-Mental State score declines occur over time. This approach will avoid the pitfall of over-diagnosing dementia in the poorly educated.

SUMMARY

The diagnosis of Alzheimer's disease is a diagnosis of exclusion. First, dementia must be properly diagnosed using the four-part definition mentioned above and considering the caveat regarding educational levels. Next, the differential diagnosis of dementia must be examined, disease by disease, with clinical and laboratory criteria used to accept or reject each diagnosis. If all other causes of dementia are rejected this way (usually after a thorough history and physical, B-12 and thyroid levels, CT scan, and removal of possibly offending medications), then a diagnosis of probable Alzheimer's disease is made. If at that time the pattern of disease progression, mental status abnormalities, and personality changes is consistent with one of the presentations of Alzheimer's, the diagnosis is made with confidence. If these patterns are not consistent, re-examining the differential diagnosis is indicated.

REFERENCES

1. McKhann, G., Drachman, D., Folstein, M. (1984). Clinical diagnosis of Alzheimer's disease: Report of the NINEDS-ADRDA Work Group under the auspices of the Department of Health and Human Services Task Force on Alzheimer's Disease. *Neurology.* 34:939-944.
2. Folstein, M. F., Folstein, S. E., McHugh, P. R. (1975). Mini-Mental State: A practical method for grading the cognitive states of patients for the clinician. *J. Psychiatr. Res.,* 12:189-198.
3. Becker, P. M., Feussner, J. R., Mulrow, C. D. (1985). The role of lumbar puncture in the evaluation of dementia: The Durham Veterans Administration/Duke University Study. *J. Am. Ger. Soc.,* 33:392-396.
4. Simon, R. P. (1985). Neurosyphilis. *Arch. Neurol.,* 42:606-613.

5. Escobar, J. I., Burnam, A., Karno, M., Forsythe, A., Landsverk, J., and Golding, J. M. (1986). Use of the Mini-Mental State Examination (MMSE) in a community population of mixed ethnicity. *J. Nerv. Ment. Dis.*, *174*:607-614.
6. Holzer, C. E., Tischler, G. L., Leaf, P. J. (1984). An epidemiologic assessment of cognitive impairment in a community population. *Res. Community Mental Health*, *4*:3-32.

CHOLINERGIC DRUG STUDIES IN DEMENTIA AND DEPRESSION

Paul A. Newhouse, M.D.

Director, Geriatric Psychiatry Service
Neuroscience Research Unit, Department of Psychiatry
University of Vermont College of Medicine
Burlington, Vermont

INTRODUCTION

For over 30 years researchers have been investigating the involvement of central cholinergic systems in the processes of memory, learning, attention and other cognitive operations. These studies received a boost when the hypothesis was generated that cholinergic lesions in the brains of patients suffering from Alzheimer's Disease might be related to the cognitive pathology of that disorder (Coyle et al., 1983). This hypothesis has led to a number of therapeutic studies in Alzheimer's Disease aimed at improving the cognitive symptomatology by cholinergic stimulation or replacement (for example, Brinkman et al., 1982; Davis and Mohs, 1982; Ferris et al., 1979; Mohs et al., 1985). The results of the studies have been for the most part disappointing; the reasons for this remain incompletely understood. This chapter will review a series of studies done in our laboratory at the National Institute of Mental Health attempting to examine in more detail the clinical relevance of cholinergic pathology in Alzheimer's disease. The general paradigm uses acute doses of cholinergic agonists and antagonists as pharmacological probes of the functional status of the central cholinergic system in demented patients suffering from Alzheimer's Disease, and as comparison groups, elderly normals and elderly patients suffering from depression.

The rationale for studying cholinergic drugs in Alzheimer's Disease includes the evidence for cholinergic functional deficits in the Alzheimer brain. The nucleus basalis of Meynert in the basal forebrain, which sends cholinergic projections to the cerebral cortex, has been shown to degenerate extensively in Alzheimer's Disease and shows a loss of acetylcholine containing cells (Coyle et al., 1983). The major findings include a loss of cholinergic cell bodies in the nucleus basalis with accompanying loss of cortical markers of acetylcholine including a decrease in choline acetyl transferase, the synthetic enzyme, and decreases in acetylcholinesterase, the enzyme responsible for breaking down acetylcholine (Bartus et al., 1982; Coyle et al., 1983). There also has been shown to be an important correlation between the decrease in cholinergic markers and the presence of pathologic hallmarks of Alzheimer's Disease such as plaques and tangles (Blessed et al., 1968). It has also been established that there is a correlation between the loss of cholinergic markers in the brain and the degree of cognitive dysfunction prior to death.

New Directions in Understanding Dementia and Alzheimer's Disease,
Edited by T. Zandi and R. J. Ham, Plenum Press, New York, 1990

65

SECTION A: CHOLINERGIC ANTAGONIST STUDIES

The initial study examined the effects of muscarinic cholinergic blockade. This was motivated not only by the pathologic findings cited above but also by the known effects of muscarinic cholinergic blockade on cognitive functioning. Acute muscarinic cholinergic blockade has been associated with a number of deficits in several areas of cognitive functioning (Drachman, 1977; Drachman and Leavitt, 1974). These include: 1) Acquisition: muscarinic blockade has been suggested to decrease the ability of organisms to store new information into memory and the time to learn new material appears to be increased; 2) Retrieval: recall of previously learned material appears to be unimpaired but recall of newly learned material during the time of exposure to the drug appears to be affected; 3) Attention: attention to stimuli appears to be decreased and vigilance is impaired; 4) Psychomotor: speed of performance declines.

In the early 1970s, Drachman and Leavitt (Drachman, 1977; Drachman and Leavitt, 1974) studied the effects of small doses of subcutaneous scopolamine in young normals and compared them to elderly normal controls. They showed that a small dose of subcutaneous scopolamine administered to young normals produced a degree of cognitive impairment that was similar to the performance of elderly normals at baseline. This suggested an age-related decline in cholinergic function. We hypothesized that we could extend this paradigm to elderly normals and Alzheimer's Disease patients and evaluate the functional relevance of the pathologic findings of cholinergic system dysfunction by using scopolamine as a pharmacologic probe in these patient groups.

Scopolamine was selected for this investigation because of its high potency and use in previous studies of muscarinic cholinergic blockade. The study was designed as a 3-dose, double-blind paradigm with 0.1, 0.25, or 0.5 mg of scopolamine administered IV. Cognitive testing took place 90 minutes following drug administration and bedside cognitive measures were performed at 1 and 2 hours following drug administration. The cognitive measures have been detailed elsewhere (Sunderland et al., 1987), but consisted of measures of new learning, Buschke Selective Reminding Task (Buschke, 1973), vigilance task (recognition of repeated words), knowledge memory (category retrieval), sustained motor effort (dynamometer) and a computer generated continuous performance task. Behavioral measures done at 1 and 2 hours included the Profile of Mood States (McNair et al., 1971), NIMH Self Rating Scale (Van Kammen and Murphy, 1975), a physical side effect rating scale, the Beck Depression Index and a mood visual analog scale. The blind observer also completed the Brief Psychiatric Rating Scale (Overall and Gorham, 1962) as well as visual analog scales. Ten Alzheimer's Disease subjects with a mean age of 58.8 ± 4 were involved in the study. Their average GDS (Reisberg et al., 1982) severity score was 4.0 ± 0.7, indicating a moderate degree of dementia. The elderly normal controls also numbered 10 with a mean age of 61.3 ± 11.2. There were 3 males and 7 females in each group. All patients met DSM-III criteria for primary degenerative dementia. They also fulfilled the clinical criteria for probable dementia of the Alzheimer's type established by the NINCDS-ADRDA (McKhann et al., 1984). Patients with a history of diabetes or head trauma, significant hypertension, seizure disorders, or other serious medical conditions were excluded. CT scans and clinical EEGs were performed on all patients and were used to exclude patients if the results did not conform to those expected with Alzheimer's Disease (for example, focal lesions on CT or asymmetries on the EEG). Normal controls were screened as outpatients to exclude those with significant medical and psychiatric illness, and they were also tested neuropsychologically with the same screening battery of tests as with the Alzheimer's patients and depressed patients to assure that their performance and cognitive tests fell within the normal range before entering the study. All subjects were free of psychotropic medications for at least 2 weeks prior to testing. No subjects had received chronic psychotropic medication in over a month.

Analysis of the results in the normal control group showed that the normal controls did not show sensitivity to the muscarinic blocking effects of scopolamine except at the 0.5 mg dose. At this dose there were significant changes in a number of cognitive and behavioral measures. Figure 1 shows the results of 3 cognitive tasks: vigilance, category retrieval and selective reminding.

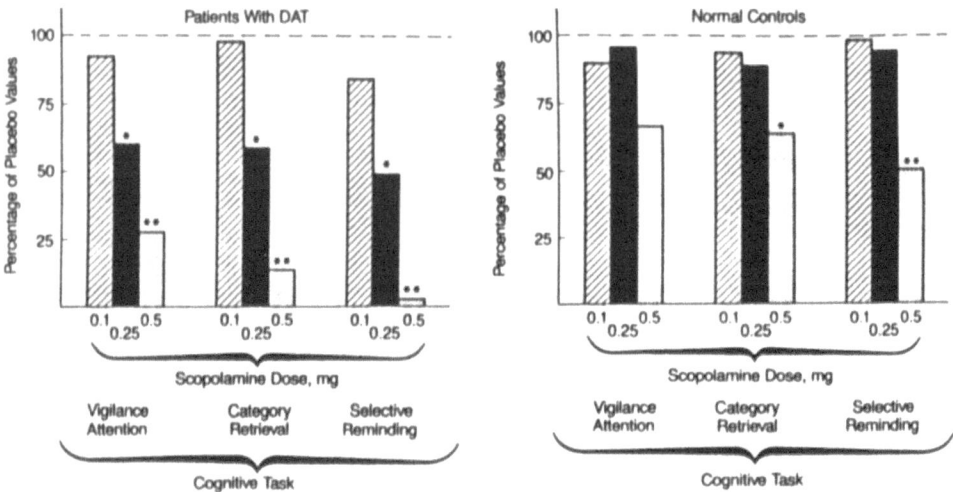

Fig 1. Cognitive testing in patients with dementia of Alzheimer type (DAT) (left, n=10) and age-matched normal controls (right, n=10) following scopolamine injection. Bars represent mean data expressed as percentages of placebo-day mean values. Asterisk indicates P<.05 compared with placebo value; double asterisk, P<.01 compared with placebo value.

Compared to the 100% placebo day baseline, after receiving 0.5 mg, the normal controls often showed a significant decrease in category retrieval and selective reminding performance and a strong trend toward the difference in the vigilance attention task. At the 0.25 mg and 0.1 mg doses however, no significant change from baseline was seen. This contrasts sharply with the results seen in the dementia patients. In Figure 1 it can be seen that compared to the performance on the placebo day, the shape of the dose response curve in the dementia patients is considerably steeper than the normal controls with the dementia subjects showing significant decline in cognitive functioning at the 0.25 mg dose as well as at the 0.5 mg dose on all three cognitive tasks. In fact, the proportion, or degree of change at the 0.25 mg dose of the dementia patients is similar to that of the 0.5 mg dose in the normal controls. These results also extended to behavioral measures as well. Figure 2 shows the Brief Psychiatric Rating Scale (BPRS) comparing both normal controls and dementia subjects and, as with cognitive tasks, the normal controls did not show sensitivity to the behavioral effects of scopolamine until the 0.5 mg dose was reached, whereas the dementia patients show a significant increase in BPRS score at the 0.25 mg dose.

Examining the results of the normal controls after receiving 0.5 mg across a number of cognitive tasks and compare them to the Alzheimer's subject (Fig 3), it can be seen that the 0.5 mg dose produces a cognitive picture in elderly normal volunteers similar to that of the Alzheimer's subjects at baseline.

This study was then extended to examine the effects of cholinergic blockade in elderly depressed patients (Newhouse et al., 1988a). Depression, particularly in the elderly, is noted to produce cognitive changes and can be of sufficient severity to be confused with a dementia process and this has been termed pseudodementia (Wells, 1979). We were therefore interested in whether elderly depressed patients would also show an increased sensitivity to muscarinic cholinergic blockade as the Alzheimer's dementia patients appear to. The same paradigm was used with 9 elderly depressed patients, 6 males and 3 females, with a mean age of 69.7 ± 6.1. Their mean severity ratings on the Hamilton Depression Rating Scale (Hamilton 1960) was 29.3 ± 7.

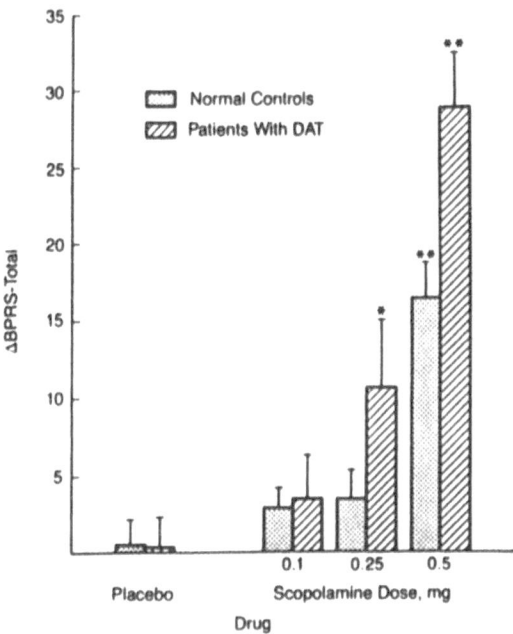

Fig 2. Mean changes in total scores on modified Brief Psychiatric Rating Scale (BPRS) across groups for each test day following placebo and scopolamine infusions. Asterisk indicates P<.05 with placebo value.

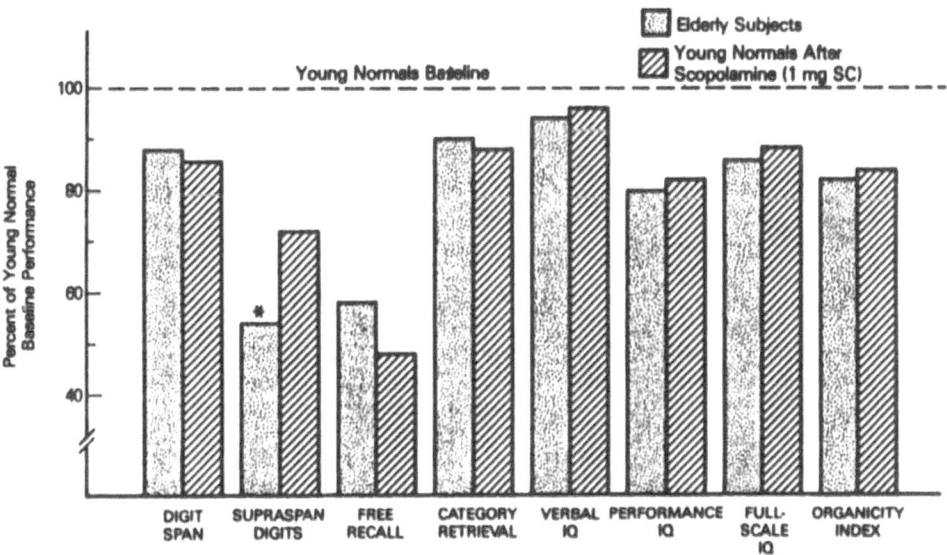

Fig 3. Comparison of normal elderly controls at baseline with young normals following administration of scopolamine (1 mg SC). Bars represent test results as a percent of the young normals' baseline scores for each task. (* P<0.05 compared to young normals after scopolamine) (Adapted from Drachman and Leavitt, 1974)

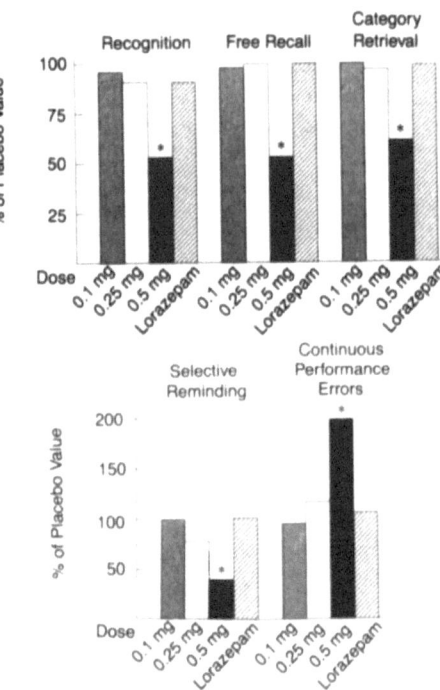

Fig. 4. Cognitive battery results expressed as percent of placebo values. Bracketed dose indicates dose of scopolamine hydrobromide; lorazepam, 1 mg orally; and asterisks, P<.05 different from placebo.

The results of the cognitive battery as seen in Figure 4, show that these subjects did not exhibit sensitivity to scopolamine except at the 0.5 mg dose.

This was a consistent picture across 5 different cognitive measures including recognition, free recall, category retrieval, selective reminding and continuous performance. A similar picture was seen with behavioral measures. There was a significant increase in BPRS score only after the 0.5 mg dose, although there was no change in depression scores. This is interesting as it has been hypothesized that cholinergic mechanisms may be involved in the modulation of affect (Janowsky et al., 1972). However, in spite of significant behavioral changes after the 0.5 mg dose, there was no significant change in the self rating of depression. These results suggest that the cognitive pathology seen in elderly depressed patients does not appear to be based on muscarinic mechanisms.

The results of the muscarinic cholinergic blockade studies tend to confirm in functional terms the relevance of the cholinergic pathology seen in the brains of Alzheimer's patients. The Alzheimer's patients show a much steeper dose response curve and increased sensitivity to muscarinic cholinergic blockade compared to both elderly normals and elderly depressed patients. Further, small doses of scopolamine in elderly normals produced a cognitive picture similar to that of dementia subjects at baseline. These results suggest that there is not only age-related but also a disease-related deterioration in cholinergic functioning. This parallels and extends the results of Drachman and others and suggests that the cholinergic pathology in Alzheimer's Disease has functional relevance to the cognitive disorder of the disease and justifies further cholinergic studies in the hope of finding ways to improve the functioning of this system.

SECTION B: MUSCARINIC CHOLINERGIC AGONIST STUDIES

A second step in understanding the cholinergic pathology of Alzheimer's disease is to examine the postsynaptic responsivity of dementia subjects to cholinergic agonists. Rather than nonspecifically challenging a system with known cholinergic abnormalities using indirect agents, we chose the postsynaptic muscarinic cholinergic agonist arecoline based on the assumption that most of the cholinergic receptors in the brain are muscarinic. For this study 12 patients suffering from Alzheimer's Disease were studied: 4 males and 8 females with a mean age of 65.8 ± 8.3 and a mean GDS score of 4.2 ± 0.7 (Tariot et al., 1988).

The 12 Alzheimer's Disease patients received infusions of placebo or choline hydrobromide at a dose of 1, 2 or 4 mg per hour of base, on 4 test days, separated by 48 hours in a randomized double-blind experiment. Glycopyrrolate 0.05 mg was administered immediately prior to the placebo and 1 mg arecoline infusion to block peripheral cholinomimetic effects. Glycopyrrolate 0.1 mg was given before the 2 and 4 mg per hour infusions. Both blood pressure and respiratory rate were also recorded at regular intervals and blood was sampled for cortisol and prolactin. Behavioral assessments included the BPRS, NIMH Self Rating Scale, and a six-item visual analog scale designed to measure cholinergic effects. A formal cognitive battery was performed at 0, 30 and 90 minutes including tests similar to those employed in the scopolamine paradigm. Speech characteristics, grammatical form, articulation and phrase length were also assessed in 9 of the subjects at 0, 30 and 90 minutes. Based on a report by Christie (Christie et al., 1981) that Alzheimer's patients showed an improved picture recognition following arecoline, a picture recognition task was also performed.

The cognitive responses of the 12 Alzheimer's Disease patients to arecoline administration are summarized in Figure 5.

Subjects did not improve in category retrieval or free recall at any of the arecoline doses. In fact there was a significant decrease in category retrieval after the 2 mg and 4 mg per hour doses. Inappropriate responses did not vary across drug conditions for either category retrieval or selective reminding tasks. Ratings of verbal expressiveness and word finding did increase significantly after 30 minutes during the 1 mg per hour infusion. There was no overall significant change on the test of picture recognition by ANOVA. But when the 4 patients who scored perfectly on this test at baseline were excluded from the analysis, a one tail t-test comparing the 2 mg per hour arecoline results with placebo in the 8 remaining Alzheimer's patients was significant at p<.05.

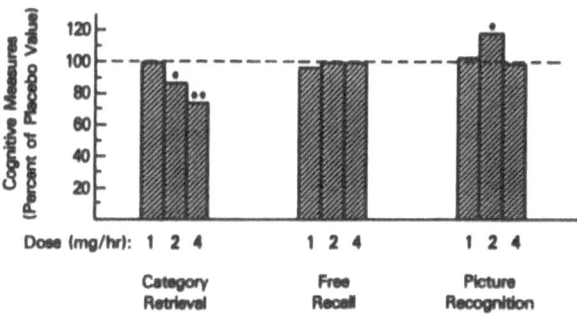

Fig. 5. Cognitive measures in Alzheimer patients (n = 12) 30 min after the start of arecoline infusions (1, 2, and 4 mg/h). Bars represent mean data expressed as percent of same-time placebo-day mean values. *P<0.05; **P<0.01.

In regard to behavioral changes, there was a significant (p<.05) decrease in the BPRS subscales of inertia (which rates emotional withdrawal, motor retardation, blunted affect and disorientation) during the 1 and 2 mg infusions compared to placebo. There was also a decrease at a trend level seen in the anxiety-depression BPRS subscale during the 1 and 2 mg per hour infusions. The visual analog affect rating showed a significant uplifting of mood at the 30-minute time point during both the 1 and 2 mg per hour infusions. The NIMH Self Rating Scale showed a significant increase in total physical symptoms only at the 4 mg per hour dose. Observer assessment of the subjects revealed mild but general activation during the 1 mg per hour infusion with increased spontaneous conversation and facial expressiveness. This was also noted during the first half hour of the 2 mg per hour infusion. During the later stages of the 2 mg per hour infusion and throughout the 4 mg per hour infusion, the patients became progressively fatigued, emotionally withdrawn and psychomotorically retarded.

The results of the muscarinic cholinergic agonist study suggested that acute cholinergic postsynaptic stimulation does not necessarily reverse the effects of the cholinergic lesions of Alzheimer's disease, however, it may be that cholinomimetics will antagonize the effect of cholinergic blockade (Sitaram et al., 1978). The sensitivity of these subjects to both the positive and negative affects of muscarinic stimulation are somewhat in contrast to previous results in normals where Sitaram and colleagues found enhanced performance on serial learning in young normals receiving subcontinuous arecoline doses of 6 mg but not with lower doses (Sitaram et al., 1978). These results and those of Christie (Christie et al., 1981) suggests that the only beneficial effects observed with arecoline occur at low doses and are consistent with the possibility that the threshold for cognitive changes is reduced in Alzheimer's disease or that the dose response curve is shifted to the left.

SECTION C: NICOTINIC CHOLINERGIC AGONIST STUDIES

The decision to examine the effects of acute nicotinic stimulation in Alzheimer's Disease was based on a number of factors. The possible beneficial effects of nicotine administration on cognitive functioning have been studied for a number of years, particularly with studies of cigarette smoking. Studies done over the previous 20 years have suggested that cigarette smoking may improve certain aspects of cognitive performance (Wesnes and Revell, 1984). There have been suggestions that nicotine or smoking may improve vigilance performance, improve information processing, reduce information processing time, and improve learning under high interference conditions (Warburton et al., 1986; Wesnes and Revell, 1984). Claims have been made that nicotine improves both short and long term recall, produces state dependent learning, and may even reverse the effect of muscarinic cholinergic blockade (Wesnes and Revell, 1984; Wesnes and Warburton, 1983). A second factor motivating these studies is the increasing recognition that there are significant numbers of central nicotinic cholinergic receptors in this central nervous system and that these receptors may play important roles in modulating neuronal function, both cholinergic and noncholinergic (Schwartz and Kellar, 1982; Shimohama et al., 1985). In addition, several groups have shown that Alzheimer's Disease subjects show a significant decrease in the number of central nicotinic binding sites (Whitehouse et al., 1986; Flynn and Mash, 1985). Although the exact localization of these receptors or binding sites is not known, they may well be associated with other cholinergic neurons in the cortex (Romano and Goldstein, 1980). Nicotine has also been reported to reverse some of the effects of central cholinergic lesions in experimental animal models (Ksir and Benson, 1983).

Therefore, we decided to examine the effects of acute nicotine administration in Alzheimer's patients (Newhouse et al., 1988). After studying young normal controls to establish tolerance, toxicity and dose parameters (Newhouse et al., 1986a), we examined 12 Alzheimer's Disease patients with a mean age of 66.8 ± 8.8 and a mean GDS rating of 4.5 ± 0.5. These subjects were studied in a single blind study where 3 doses of nicotine bitartrate and placebo were used. Doses were 0.125, 0.25 and 0.5 µg/kg/minute of nicotine base administered for 60 minutes. Behavioral ratings included the BPRS, NIMH Self Rating Scale and a 14 item observer visual analog scale, similar to the scale used in the scopolamine testing; these were completed at baseline, 30, and 60 minutes. A cognitive battery was also administered at 0, 30, 60 minutes, and at 4, 8 and 24 hours; tests on this battery included tests of free recall, and category retrieval. At the 4, 8 and 24 hour sessions, only recall was tested.

Plasma nicotine levels in normal volunteers showed that the 0.5 µg dose produced a level at 60 minutes of approximately 16 ng/ml of nicotine. This is roughly 1/2 to 1/3 of the plasma nicotine seen in smokers after receiving a single cigarette. Results of cognitive testing showed that correct free recall did not exhibit any significant change across the doses. Total word production was essentially flat across the 4 doses.

However the intrusion errors did show significant (p<.05) change. In Figure 6 it can be seen that the intrusion errors declined significantly after the 0.25 µg dose at both 30 and 60 minutes into the infusion. The curve exhibits the characteristics of a U-shaped dose response curve, with improvement at a middle dose and less improvement or no improvement at both lower and higher doses. That this was not due to simply a reduction in total word production can be seen by the finding that total free recall was not significantly different across the 4 conditions. In fact, total words produced were not significantly different across any of the doses. Examination of long-term memory showed that there was a significant improvement in recall consistency at the 8-hour testing period. Figure 7 shows the recall consistency at 8 hours from each of the word lists administered at 30 and 60 minutes.

After receiving the 0.25 µg dose, subjects were more likely to remember words at 8 hours than they had initially remembered at the time the list was administered. This suggests that there is some evidence that nicotine improves long-term memory or memory consolidation.

These cognitive results, however, were not without behavioral effects. Depressive affect and anxiety self-ratings showed significant increases particularly after the 0.5 µg dose. Anxiety ratings showed a consistent picture across multiple measures including significant increases in subjective anxiety visual analog scale, the NIMH Self Rating Anxiety Subscale and observer ratings of anxiety. These mood and anxiety affects in some patients were clinically quite dramatic. For example, a 54-year-old man with Alzheimer's disease who had no prior history of an anxiety disorder experienced a panic attack during the 0.5 µg infusion of nicotine (Newhouse et al., 1986b). In several other instances, patients became depressed and tearful during the

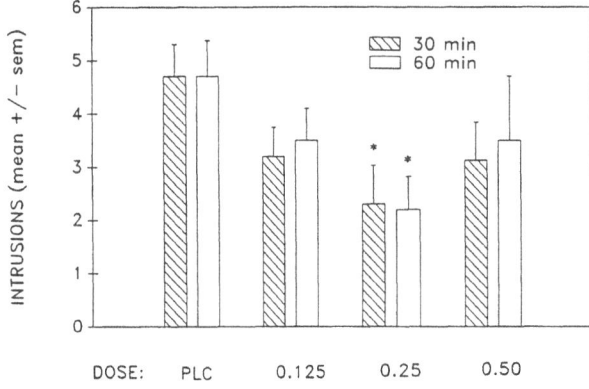

Fig. 6. Intrusion errors after intravenous nicotine in Alzheimer's disease patients at 30 and 60 minutes after beginning of infusion. Dose expressed as µg/kg/min of nicotine base. *P<.05 different from placebo.

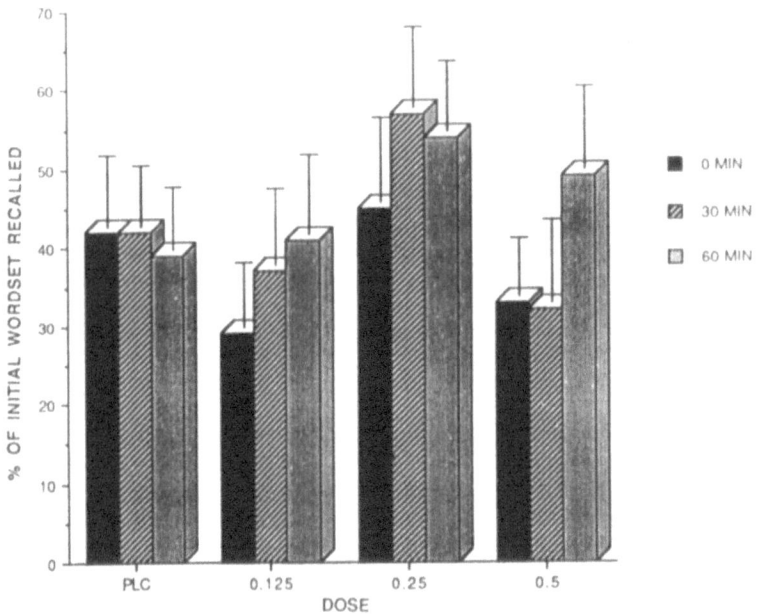

Fig. 7. Recall consistency at 8 hours after infusion of nicotine for words initially presented at 0 (predrug), 30 and 60 minutes (during infusion). Dose expressed as μg/kg/min of nicotine basis.

0.5 μg infusion without experiencing concomitant significant physical side effects. Normal volunteers in general did show some dose-related increase in anxiety but the magnitude of the change was not as great with the patient subjects. It should be noted though that these responses were somewhat sporadic in that some patients became highly anxious and other patients did not show significant anxiety changes.

There were significant (p<.05) dose related changes in physiological parameters such as systolic blood pressure, pulse, and respiratory rate. The total number of physical side effects on the NIMH scale was found to be significantly increased at the highest dose. The most common physical side effects were mild headaches, dizziness, and sense of unease. Measurements of endocrine parameters showed a significant (p<.01) dose-related increase in plasma ACTH, cortisol and prolactin, which were consistent with central nicotinic stimulation.

In summary, the effects of intravenous nicotine administration in this pilot study were to show evidence for improvement in information selectivity indicated by the decline in intrusion errors on the free recall task and improvement in long-term memory stores or memory consolidation suggested by the improvement by 8-hour recall consistency. Behavioral results showed a significant dose related increase in negative affect and anxiety which may have compromised improvement in cognitive functioning, particularly at the 0.5 μg dose. Dose-related physiologic and endocrine changes were also seen consistent with central nicotinic stimulation. These results suggest that there may be some potential in further exploration in nicotinic stimulation in Alzheimer's Disease and more careful examination of the functional relevance of the established nicotinic receptor pathology in this disorder.

<u>Comments</u>

These studies have attempted to take a systematic approach to the evaluation of the functional significance of the observed cholinergic system lesions in Alzheimer's Disease by using a variety of pharmacological probes to test the function and responsiveness and sensitivity of central cholinergic neurons in Alzheimer's Disease and in the elderly in general. These studies have a number of limitations, not the least of which being the limited numbers of control subjects in the 2 agonist studies and the lack of a similar antagonist study with a nicotinic blocker. Further, the adverse behavioral effects of the 2 agonists drugs preclude firm conclusions being made about the therapeutic potential of cholinergic augmentation. It has been suggested that agonist type drugs may not be ideal for long-term use in Alzheimer's Disease as they do not simulate the normal physiological traffic across neuronal synapses but rather over stimulate both pre- and post-synaptic receptors. Further, in the case of nicotine there is evidence that chronic nicotine administration may produce desensitization of central nicotinic receptors and up-regulate receptor numbers, probably secondary to this desensitization (Marks et al., 1987). This may limit the usefulness of nicotinic administration unless ways can be found to limit this progressive desensitization and receptor upregulation. The function of the nicotinic receptors that have been found are not precisely known nor is their location firmly established. They may be presynaptic in nature and modulate the release of acetylcholine or other neurotransmitters in the cortex, and there may be some on post-synaptic neurons as well. Whether nicotinic stimulation may positively affect other neurotransmitters systems remains to be seen.

The scopolamine studies established that muscarinic cholinergic systems clearly must be involved in the cognitive disorder of Alzheimer's Disease although it is not entirely clear exactly what types of cognitive operations these systems modulate. It is possible that there are integrated nicotinic and muscarinic systems in the cortex where nicotinic receptors modulate acetylcholine release onto postsynaptic muscarinic receptors. Cholinergic cell loss and the loss of cholinergic markers in the cortex have also been described in other degenerative neurologic disorders including Parkinson's Disease and progressive supranuclear palsy (Whitehouse et al., 1983). Nicotinic receptor loss has also been described in these disorders (Perry et al., 1987). It is possible therefore that there is some unitary set of cognitive operations that may be affected by cholinergic systems in all 3 disorders but precise analysis of these cognitive operations remains to be worked out. It is possible that cholinergic systems may be more involved in some types of cognitive operations than others. Further studies will be required to establish what these are.

Implications for therapy from these studies indicate the difficulty in using the cholinergic agonists that are currently available. New delivery technology such as skin patch administration or less toxic and more specific agonists compounds will be required before cholinergic agonists can be used as a practical therapeutic strategy in dementia.

The author would like to thank Drs. Pierre Tariot and Trey Sunderland for permission to cite their work.

REFERENCES

Bartus, R. T., Dean, R. L., Beer, B. and Lippa, A. S. (1982). The cholinergic hypothesis of geriatric memory dysfunction, *Science, 217*:401-417.
Blessed G., Tomlison, B. E., Roth, M. (1968). The association between quantitative measures of dementia and of senile change on the cerebral grey matter of elderly subjects. *Br. J. Psychiatry, 114*:797.

Brinkman, S. D., Pomara, N., Goodnick, P. J., Barnett, M. A. and Domino, E. F. (1982). A dose-ranging study of lecithin in the treatment of primary degenerative dementia (Alzheimer disease), *J. Clin. Psychopharmacol.*, *2*:281-285.

Buschke, H. (1973). Selective reminding for analysis of memory and learning. *J. Verb. Learn. Behav.*, *12*:543-550.

Christie, J. E., Shering, A., Ferguson, J. and Glen, A. I. M. (1981). Physostigmine and arecoline: Effects of intravenous infusions in Alzheimer presenile dementia, *Br. J. Psychiatry*, *138*:46-50.

Coyle, J. T., Price, D. L. and DeLong, M. R. (1983). Alzheimer's disease: a disorder of cholinergic innervation. *Science*, *219*:1184-1190.

Davis, K. L. and Mohs, R. C. (1982). Enhancement of memory processes in Alzheimer's disease with multiple-dose intravenous physostigmine. *Am. J. Psychiatry*, *139*:1421-1423.

Drachman, D. A. (1977). Memory and cognitive function in man: Does the cholinergic system have a specific role? *Neurology*, *27*:783-790.

Drachman, D. A. and Leavitt, J. (1974). Human memory and the cholinergic system. *Arch. Neurol.*, *30*:113-121.

Ferris, S. H., Sathananthan, G., Reisberg, B. and Gershon, S. (1979). Long-term choline treatment of memory-impaired elderly patients. *Science*, *205*:1039-1040.

Flynn, D. D. and Mash, D. C. (1985). Nicotine receptors in human frontal add infratemporal cortex: Comparison between Alzheimer's disease and the normal. *Neurosci. Abstr.*, *11*:1119.

Hamilton, M. A. (1960). A rating scale for depression. *J. Neurol. Neurosurg. Psychiatry*, *23*:56.

Janowsky, D. S., El-Yousef, M. K., Davis, J. M. and Sekerke, H. J. (1972). A cholinergic-adrenergic hypothesis of mania and depression. *Lancet*, *1*:632-635.

Ksir, C., Benson, D. M. (1983). Enhanced behavioral response to nicotine in an animal model of Alzheimer's disease. *Psychopharmacology*, *81*:272-273.

Marks, M. J., Stitzel, J. A., Collins, A. C. (1987). Influence of kinetics of nicotine administration on tolerance development and receptor level. *Pharm. Biochem. Behav.*, *27*:505-512.

McKhann, G., Drachman, D., Folstein, M., Katzman, R., Price, D. and Stadlan, E. M. (1984). Clinical diagnosis of Alzheimer's disease: Report of the NINCDS-ADRDA work group. *Neurology*, *34*:939-944.

McNair, D. M., Loor, M. and Droppleman, L. F. (1971). *Profile of Mood States*. Educational and Industrial Testing Service, San Diego, CA.

Mohs, R. C., Davis, B. M., Johns, C. A., Mathe, A. A., Greenwald, B. S., Harvath, T. B. and Davis, K. L. (1985). Oral physostigmine treatment of patients with Alzheimer's disease. *Am. J. Psychiatry*, *142*: 28-33.

Newhouse, P. A., Sunderland, T., Thompson, K., Tariot, P. N., Weingartner, H., Murphy, D. L. (1986a). Dose related physiologic, behavioral, and cognitive effects of nicotine on naive human subjects. *Neurosci. Abstr.*, *12*:1445.

Newhouse, P. A., Sunderland, T., Thompson, K., Tariot, P. N., Weingartner, H., Mueller, E. R., Cohen, R. M., Murphy, D. L. (1986b). Intravenous nicotine in a patient with Alzheimer's disease. *Am. J. Psychiatry*, *143*:1494-1495.

Newhouse, P., Sunderland, T., Tariot, P., Thompson, K., Weingartner, H., Mellow, A., Cohen, R. M., Murphy, D. L. (1988). The effects of acute scopolamine in geriatric depression. *Arch. Gen. Psych.*, *45*:906-912.

Overall, J. E. and Gorham, D. R. (1962a). The Brief Psychiatric Rating Scale, *Psychol. Rep.*, *10*:799-812.

Perry, E. K., Perry, R. H., Smith, C. J., Dick, D. J., Candy, J. M., Edwardson, J. A., Fairbairn, A., Blessed, G. (1987). Nicotinic receptor abnormalities in Alzheimer's and Parkinson's disease, *J. Neurol. Neurosurg.*, *50*:806-809.

Reisberg, B., Ferris, S. H., DeLean, M. J. and Crook, T. (1982). The global deterioration scale for assessment of primary degenerative dementia, *Am. J. Psychiatry, 139*:1136-1139.

Rowell, P. P., Winkler, D. L. (1984). Nicotinic stimulation of [3H] acetylcholine release from mouse cortical synaptosomes. *J. Neurochem., 43*:1593-1598.

Schwartz, R. D., Kellar, K. J. (1982). Nicotinic cholinergic receptors labeled by [3H]-acetylcholine in rat brain. *Mol. Pharmacol., 22*:55-62.

Shimohama, S., Taniguchi, T., Fujiwara, M., Kameyama, M. (1985). Biochemical characterization of the nicotinic cholinergic receptors in human brain: Binding of (-)-[3H] nicotine. *J. Neurochem, 45*:604-610.

Sitaram, N., Weingartner, H. and Gillin, J. C. (1978). Human serial learning: Enhancement with arecoline and choline and impairment with scopolamine. *Science, 201*:274-276.

Sunderland, T., Tariot, P. N., Cohen, R. M., Weingartner, H., Mueller, E. A. and Murphy, D. L. (1987). Anticholinergic sensitivity in patients with dementia of the Alzheimer type and age matched controls: A dose-response study. *Arch. Gen. Psychiatry, 44*:418-426.

Tariot, P. N., Cohen, R. M., Welkowitz, J. A., Sunderland, T., Newhouse, P. A., Murphy, D. L. and Weingartner, H. (1988). Multiple dose arecoline infusions in Alzheimer's disease. *Arch. Gen. Psychiatry, 95*:901-905.

Van Kammen, D. P. and Murphy, D. L. (1975). Attenuation of the euphoriant and activating effects of d- and l-amphetamine by lithium carbonate treatment. *Psychopharmacologia, 44*:215-224.

Warburton, D. M., Wesnes, K., Shergold, K., James, M. (1986). Facilitation of learning and state dependency with nicotine. *Psychopharmacology, 89*: 55-59.

Wells, C. E. (1979). Pseudodementia. *Am. J. Psychiatry, 136*:895-900.

Wesnes, K., Revell, A. (1984). The separate and combined effects of scopolamine and nicotine on human information processing. *Psychopharmacology, 84*:5-11.

Wesnes, K., Warburton, D. M. (1983). Smoking, nicotine, and human performance. *Pharmacol Ther., 21*:189-208.

Whitehouse, P. J., Hedreen, J. C., White, C. L., Price, D. L. (1983). Basal forebrain neurons in dementia of Parkinson's disease. *Ann. Neurol., 13*:243-248.

Whitehouse, P. J., Martino, A. M., Antuono, P. G., Lowenstein, P. R., Coyle, J. T., Price, D. L., Kellar, K. J. (1986). Nicotinic acetylcholine binding sites in Alzheimer's disease. *Brain Res., 371*:146-151.

Section II

Cognitive and Language Evaluations of Dementia Patients

VERBAL COMMUNICATION IMPAIRMENT IN DEMENTIA

RESEARCH FRONTIERS IN LANGUAGE AND COGNITION

Raymond A. Domenico, Ph.D.

Chair, Department of Hearing and Speech Science
State University of New York, Plattsburgh

A careful study of language changes in Alzheimer's can be useful in at least three areas: (1) in identification and differential diagnosis, (2) in formulating intervention strategies and (3) in the study of neurolinguistics and the relationship between language and thought. Specifically, Dementia of the Alzheimer's type (DAT), as in all other progressive as well as reversible dementias, is characterized by a linguistic disorder as well as a profound communication disorder (Overman & Geoffrey, 1987; Bayles, 1985; Hier, Hagenlocker & Shindler, 1985). The linguistic and cognitive bases of language in comparison with the communication aspect of language are important concepts in understanding the disruptive effects of this disease and warrant explication. Thus, this chapter will attempt to lead the reader to a greater understanding of the central role which language plays in understanding and treating this disorder, and, how the study of this disorder may offer insight into the interdependence of language and thought.

COMMUNICATION AND LANGUAGE

Communication can be thought of as the sending and receiving of messages through verbal or nonverbal means. At this simple level, all that is required for communication to occur is that there be some form of interaction between participants and that messages be exchanged. The communication act does not even necessarily depend on linguistic abilities or what we commonly think of as language (Martin, 1981). However, when the message requires the exchange of significant amounts of accurate information or the communication of abstract ideas, then we rely on our language ability (cognitive related) for the processing and sending of information, and we most typically use a verbal (or written) communication mode. Moreover, verbal communication, at the cognitive level, relies on specific linguistic or language subsystems referred to as: (1) semantics, the knowledge of the meanings of individual words and word combinations; (2) syntax, the knowledge of the rules of word order; (3) phonology, the knowledge of the system of sounds and rules for combining them; (4) morphology, the knowledge of the structure of words and how elements such as suffixes can be combined with root words to form new words; and (5) pragmatics, the knowledge of when and how to use language for such things as turn taking in conversation, determining the intent of a speaker, or in using language to persuade, intimidate or deceive a listener (Bayles, 1985).

New Directions in Understanding Dementia and Alzheimer's Disease,
Edited by T. Zandi and R. J. Ham, Plenum Press, New York, 1990

79

LANGUAGE AND COGNITION

Operation of these linguistic subsystems depends on the action and interaction of underlying cognitive processes by which sensory information is received, elaborated, stored, retrieved and used. In other words, language is supported by such cognitive functions as short-term memory, word retrieval, feature abstraction and other mental processes (Martin, 1981). Thus, cognition as it supports language, allows us to comprehend and express subtle and abstract thoughts through a complex and interactive process which ultimately links sound and meaning.

While there are disturbances in orientation, memory, cognition and personality, it is the cognitive impairment which is central to DAT and its consequences. Simply put, the ability to think consciously (and later subconsciously), to reason, to mentally manipulate and synthesize information and to learn is insidiously and severely compromised over the course of the disease. It is not surprising, therefore, that the language subsystems which are dependent on a deteriorating hierarchy of cognitive processes will be among the first to show change (Hier, Hagenlocker & Shindler, 1985; Appell, Ketesz & Fisman, 1982; Bayles, 1985).

LANGUAGE CHANGES IN DAT

Changes in language and speech in dementias, particularly in DAT have been documented and indicate a fairly predictable progression from loss of vocabulary and naming to deficits in social-communicative pragmatics to inadequate syntax for expressing abstractions, to increasing instances of jargon/nonsense speech to logorhea, empty speech and, finally, mutism (Overman & Geoffrey, 1987). A more detailed analysis of the progressive changes in language in DAT are given in Table 1 and Table 4 (Overman & Geoffrey, 1987). Here, changes in various linguistic skills (naming, verbal expression, comprehension, repetition, automatic speech, writing, reading) which parallel the disease's progress from the initial-mild stage through to the late-severe stage are given. This kind of detailed breakdown across several language modalities over the course of the disease is helpful in identification of DAT and in differential diagnosis.

ROLE OF LANGUAGE CHANGES AND DIFFERENTIAL DIAGNOSIS

While linguistic changes are helpful in differential diagnosis, other indicators have been used when comparing Alzheimer's disease to other neurological and neuro-psychiatric disorders such as Pick's disease, multi-infarct dementia (MID), depression and various aphasic syndromes (Cummings, 1985). Table 2 summarizes these linguistic and non-linguistic discriminators described in several studies (Cummings, 1985; Golper & Binder, 1981; Appell, Ketesz & Fisman, 1982; Hier et al., 1985).

Table 2 shows how the disorders listed differ from Alzheimer's disease. As such, any description of changes in other dementias listed should be seen in contrast to the changes noted in DAT. For example, the linguistic discriminator between DAT and depression suggests that while the use of semantic cues helps those with depression, their use does not help those with DAT (Golper & Binder, 1981).

DAT AND APHASIA

Recent research comparing DAT and aphasia suggests that the similarities and differences are much more complex than the initial work in this area had indicated. Specifically it was once thought that the differences between the two diseases was definitive. Now this thinking has changed. In fact, when one compares the various aphasia syndromes to the language disorders seen in DAT as several studies have done (Appell et al., 1982; Gewirth, Shindler & Hier, 1984; Nicholas, Obler, Albert & Helm-Estabrooks, 1982 & 1985), it appears that the language disorders of DAT resemble different aphasia syndromes as the disease progresses. Hier et al. (1985), have tabulated some of these comparisons (see Table 3).

TABLE 1

Linguistic Characteristics of Alzheimer's Disease (AD)
Mild to Moderate Stage

Linguistic Function	*Characteristics*
Early or Mild Stage of DAT	
Naming	Some errors noted on naming tasks; substantial number of pauses to search for correct word; may substitute word not correct and will then often correct error.
Verbal Expression	Discourse is usually intact and informative although at times overly long.
Comprehension	Can do quite well in attending to single-sentence questions and sentences with complex syntax. Difficulty with story-level material.
Repetition	Can repeat more frequently occurring words in sentences unless length is too long; some problems begin to show in repeating less frequently occurring words.
Automatic Speech	Can initiate series on own, may run on after end or may omit one or two items.
Writing	Can usually write a paragraph; some errors in spelling.
Reading	Can read aloud, may repeat words and produces few errors, comprehension diminishes as material increases in length and complexity.
Middle or Moderate Stage of DAT	
Naming	Many errors on naming tasks errors include circumlocution, and both semantic and literal paraphasias; semantic or sound cues do not help aphasia.
Verbal Expression	Includes paragramatisms and clang association, cannot tell a coherent story.
Comprehension	Cannot answer questions asked; appear to be answering something else; on structured tasks does moderately well, but has more problems as length and complexity increase.
Repetition	Has difficulty with all less frequent occurring words when utterance length increases beyond six words.
Automatic Speech	Needs prompting to begin a series, often has problems completing it or omits items either within or at end of series.
Writing	Can produce no more than one sentence, may be able to write names, writing contains spelling errors and grammar may be impaired; does not recognize errors.
Reading	In reading aloud substitutes one word for another or produces a nonsense word, cannot read paragraph length material or answer questions.

Note: From "Alzheimer's Disease and Other Dementias" by C.A. Overman and V.C. Geoffrey. In *Communications Disorders in Aging: Assessment and Management* (p. 274) by H.G. Mueller and V.C. Geoffrey (Eds.), 1987, Washington: Gallaudet University Press. Copyright 1987 by Gallaudet University. Adapted by permission.

It appears that in the early-mild stages of the disease, patients with DAT have primarily a naming disorder and sound most like an anomic aphasic. As the disease progresses, their language behavior resembles a Wernicke's aphasia. This does not indicate, however, that patients with DAT have a true aphasia in the classical sense. In fact, determining if the language disorder in DAT should be considered an aphasia has become a controversial issue (Davis, 1983). This controversy was evident at the 1988 annual convention of the American Speech Language and Hearing Association (Asha) where a mini-seminar and panel discussion/debate was conducted by leaders in speech-language pathology, neurolinguistics, and neuropsychology on the topic "Are the communication disorders of dementia appropriately characterized as aphasia?" (Bayles, Tomoeda, Obler, Albert, Wertz, Au and Gonzales-Rothi, 1988). Apparently some researchers use the term aphasia without regard for the classical body of knowledge which has accrued over time and which defines the parameters of this condition with respect to etiology, course and other distinctive elements (Davis, 1983).

It is important to clarify whether or not DAT may be classified as an aphasia for several reasons including the development of appropriate intervention strategies and gaining a better understanding of brain-language relationships which we will discuss below.

For now, suffice it to say that (as indicated in Table 2) Wernicke's aphasics while, perhaps, not comprehending the specific meanings of words and utterances any better than the patient with advanced DAT, do understand the intent of the utterance, where those with DAT do not (Albert, 1981). There are some other linguistic differences, but the lack of perceived intent seems to be a most exquisite indicator that the patient with DAT primarily has a problem with thinking which secondarily affects language. This suggests that the impairment is diffuse, affecting such non-primary linguistic areas of the brain as the right hemisphere which provides the aphasic with a level of thinking (although subconscious) and understanding (i.e., being tuned-in) not available to the patient with advanced DAT.

INTERVENTION STRATEGIES

As mentioned above, one important reason to clarify if the language disorder of DAT is an aphasia has to do with developing appropriate intervention strategies. This population, when seen in treatment, is often provided techniques which have been developed for the treatment of aphasia. These rehabilitation approaches include non-specific approaches such as environmental stimulation and socialization groups, to specific approaches such as: (1) stimulation facilitation, (2) programmed operant therapy, and (3) deblocking techniques (Davis, 1983).

Stimulation facilitation relies on intensive auditory stimulation and strives for maximum response. The operant approach is a traditional behavior modification approach with baseline measures and operational definitions of terminal/target behaviors. Deblocking techniques are used to stimulate central language processes by using a spared channel, such as hearing a series of words including the target word. This is simply a priming technique which has been used in studying normal language processing and language in aphasia and dementia.

Without modifications these direct language treatment techniques have not met with great success when used in DAT (Davis & Baggs, 1984; Overman & Geoffrey, 1987) although more study is needed. Modified reality orientation approaches, however, while not showing lasting effects, can be used to help keep patients engaged in their environment. Through engagement and increased awareness of their environment the connections between communicative attempts and the communicative context are made more apparent. Such methods also increase opportunity for affiliation and foster a sense of belonging within the patient.

One such program is currently being conducted at the Hebrew Home for the Aged in Riverdale, New York. This technique labelled "Dialogue and Discourse" (Witte, 1989)

TABLE 2

Linguistic Behavior of Dementia of Alzheimer's Type
Compared With Other Forms of Dementia,
Depression and Aphasia

DAT	PICK'S	MID	DEP	APHASIAS
		Linguistic Differences		
Initially fluent speech with preserved syntax, but difficulty in naming not helped by cueing. Naming errors not target related. Obvious loss of pragmatic skills. Language loss secondary to cognitive disorder.	Almost identical to DAT. More pronounced logoclonia* in late stage.	Uses fewer words poorer syntax. Telegraphic verbal expression.	Use of semantic cues help naming. skills.	Nonfluent or show more neologisms** early. These are target related. Better semantic and pragmatic Primarily a language disorder.
		Other Cognitive Differences		
Early loss of memory functions and other higher mental processes. Disoriented but retains mechanics of reading and writing until late stage.	Retains memory visuaospatial and arithmetic skills longer than DAT.	Intact nonverbal memory, oriented, no reading aloud.	Better long-term memory. Variable cognitive ability.	Oriented, good memory. Poorer on reading and writing.
		Neurological Differences		
Diffuse cortical pathology with most damage concentrated in tempero-parieto-occipital junction.	Lobar atrophy. Diffuse damage concentrated in Frontal/Temporal area. Pick's Bodies	Multiple focal lesions. Step-wise course.	Normal EEG EEG & CAT Scan.	Discrete areas of damage in language areas; Frontal, Temporal, Parietal Lobes
		Other Differences		
Progressive course often with no motor impairment but change in personality with eventual complete	Early signs of Kluver-Bucy flamboyance. Seizures are rare.	Abrupt onset. Hx. of C.V.A. Focal signs. Personality preserved.	Vegetative signs. Psychomotor retardation, affect depressed. Self limiting.	Non-cumulative. Often motor impairment but self-awareness preserved.

MID = multinfarct dementia
DEP = depression with dementia
* logoclonia (def) The continual repetition of a word or sounds, apparently without meaning.
** Neologism (def) An unintentionally produced nonsense word form not found in the lexicon of a language.

helps patients in the latter stages of DAT to engage in the pragmatics of communication through a structured group intervention approach. These patients often do not initiate speech or use full sentences and cannot maintain a conversation. However, when given communication support from a structured group setting, they retain a sense of participation and perhaps can be helped to retain certain general communication functions longer.

According to Witte (1989) this approach is not intended to promote new learning or to restimulate or reorganize language function as is expected in aphasia therapy. Still, in terms of the linguistic vs. communication distinction mentioned at the beginning of this chapter, this technique suggests a communication approach to intervention to be most effective.

Other suggestions from Witte (1989) for communication which may be helpful to those working with patients with DAT are:

1. Pair concrete objects and visual cues with verbal messages.

2. Speak somewhat louder and slower.

3. Speak on topics of interest to the patient.

4. Do not switch topics quickly or without warning.

5. Use gesture paired with short simple sentences.

6. Avoid detailed descriptions/use general terms.

7. Watch for non-verbal signs.

8. Do not argue a point. State the idea differently.

9. Try repeating questions and comments.

10. Label the environment with pictures and words.

ANOMIA IN DAT AND APHASIA

While methods such as those mentioned, address the issue of facilitating communication in general and point to communicative differences between aphasics and patients with DAT, much of the debate remains at the subordinate linguistic level. Specifically, the question as to whether the language disorder in DAT should be considered aphasia relates to the issue of the exact nature of the naming problems or anomia seen in both disorders. Is the problem one of a <u>dissolution of the memory</u> of the meanings of words, that is, a semantic memory problem? (This was postulated by Bayles, Tomoeda, Kaszniak, Stern & Eagans, 1985.) Or is the problem one of <u>accessing</u> semantic memory as has been postulated concerning the aphasias?

TABLE 3

Similarities Between Dementia Group Means
and Aphasia Group Means

Test	Early SDAT	Late SDAT	Early SRD	Late SRD
MLU	Anomic	Anomic	Anomic	Wernicke
Subordinate Clauses	Anomic	Anomic	Wernicke	Broca
Total Words	Wernicke	Anomic	Wernicke	Broca
Prepositional Phrases	Wernicke	Anomic	Wernicke	Broca
Anomia Index	Anomic	Wernicke	Normal	Normal
Empty Words	Anomic	Wernicke	Anomic	Wernicke
Conciseness Index	Anomic/ normal	Wernicke	Anomic/ normal	Anomic/ normal

From: "Language Disintegration in Dementia: Effects of Etiology and Severity" by D.B. Hier, K. Hagenlocker, and A.G. Shindler (1985) <u>Brain and Language</u>, <u>25</u>, 127. Copyright 1985 by Academic Press Inc. Adapted by permission.

SDAT -- Senile Dementia of the Alzheimer's type
SRD -- Stroke related dementia

Support for the former viewpoint was suggested by Bayles et al. (1985), who indicated that while patients with DAT can recognize and correct phonologically and syntactically incorrect sentences, they cannot or do not correct semantically inaccurate and absurd sentences; nor can they disambiguate sentences. According to Bayles et al. (1985), they retain the automatic algorithms of the lower brain centers supporting sentence structure (i.e., syntactic rules), but comprehending or expressing meaning requires conscious effort and the support of the higher brain centers.

As dementia progresses, the patient's discourse becomes increasingly empty or devoid of meaning. An example of this in a person with moderate DAT is given by Bayles (1982) in the following:

When asked to tell everything they could about a gray button, the patient said: "Oh, that's a needle. But.... Buttonhole, scissors. And they go ahead. They put buttons or they put...That's how they put buttons on your coat with it I guess."

A person with severe DAT when asked the same question about a marble, replied: "Well, he was standing there looking you know, so I. It's not mine. I didn't have it (p. 65)."

These two examples point out the anomia and loss of meaning, although in the first example it might appear that the person produced some related responses (i.e., needle and buttonhole). It might be contended that the idea or thought of button or marble is lost even though the person may be able to repeat the word "button." Simply saying the word, moreover, does not represent a semantic skill. Naming an object defined, however, or producing an antonym for a pictured word does.

Nebes (1985), on the other hand, disagrees with those who believe that this sort of response represents a loss of semantic memory. He postulates an access problem and uses the semantic priming techniques mentioned above in the discussion on deblocking to show that while they take longer than normals, patients with DAT can take advantage of semantic cueing or priming in a naming task (Nebes, 1985). To quote Nebes on this important finding:

The presence of normal semantic priming in demented patients suggests
that the structure of the associative network relating the various concepts
in their semantic memory is at least grossly intact, and that there is in
demented individuals an automatic spread of activation along this network
similar to that found in normal young and elderly individuals. Thus, while
demented patients may not be able to describe how a table and a chair are
alike, this semantic relation does still exist in the patient's memory, and
it can act to automatically facilitate (i.e., prime) processing (p. 115).

Nebes (1985) goes on to indicate that more work should be done in determining the best kinds of primers to use and that it does seem clear that semantic priming can facilitate language processing as long as "a conscious effortful search or decision is not required" (p.116).

NEUROLOGICAL SUPPORT FOR LANGUAGE AND THOUGHT

Perhaps the discrepancies between studies which seem to support one theory or the other are due to differences in the amount of damage across subjects tested. For example, it is generally believed that aphasia represents an accessing problem (where word meanings are not lost but are retrieved with difficulty). There is some evidence that when enough brain damage occurs, even in a discrete area, the function supported by that area may not be retrieved. For example, Naeser, Helm-Estabrooks, Haes, Auerbach and Scrinivasan (1987) found that when more than half the cells in Wernicke's area in the temporal lobe were damaged, auditory comprehension was severely and permanently compromised. A vast range of differences in cognitive-linguistic functioning for the early or middle stage may be seen (e.g., Table 1) compared with the late or severe stage (e.g., Table 4) where subjects cannot even attend to the task.

TABLE 4

Linguistic Characteristics of Alzheimer's Disease (AD)
Late or Severe Stage

Late or Severe Stage DAT

Naming	Severely impaired naming or inability to perform naming tasks at all.
Verbal Expression	Often does not respond; may use jargon, rambling, and/or incoherent speech.
Comprehension	Appears not to comprehend what is said.
Repetition	Unable to attend to structured test, appears to repeat what is said with clang association.
Automatic Speech	Unable to attend to structured test.
Writing	Unable to attend to structured test.
Reading	Unable to attend to structured test.

Note: From "Alzheimer's Disease and Other Dementias" by C.A. Overman and V.C. Geoffrey. In *Communications Disorders in Aging: Assessment and Management* (p. 274) by H.G. Mueller and V.C. Geoffrey (Eds.), 1987, Washington: Gallaudet University Press. Copyright 1987 by Gallaudet University. Adapted by permission.

The study of dementia reveals the interrelationship between language and thought. Both require the storage, retrieval and manipulation of symbols representing sensory information gathered through various receptors. When information processing breaks down it will affect both storage as well as access in both thinking and language.

Is there any physiological support for this concurrence of disturbances in both language and thought? The answer may be beyond us at this point in the study of neurolinguistics. However, it is interesting to note that the early Geschwind connectionistic model of language processing located the naming function to the point where the auditory, visual and somesthetic sensory association areas of the temporal, occipital and parietal lobes converge in the inferior parietal lobe (often referred to as the supramarginal and angular gyri).

This is the same area referred to by Cummings (1985) in his description of the localization of the neuropathology in DAT. Cummings states:

> The principal pathological changes in DAT include neurofibrillary tangles, senile plaques, granulovacuolar degeneration, and loss of neurons. The changes are most dense in the posterior temporal-parietal-occipital junction area, involve the anterior temporal and frontal convexity to a lesser extent, and are least prominent in the primary motor and sensory areas and orbitofrontal cortex (p. 55).

To conclude, recent investigations of language and communication impairment in Alzheimer's Disease are refining our knowledge of differential diagnostic criteria. These investigations provide us with certain implications concerning future directions for research. What is now needed is a comprehensive longitudinal study of language change in DAT with multiple neurological and neuro-psychological measures.

Intervention strategies must continue to be attempted despite the progressive nature of the disorder. Determining the most effective priming methods, perhaps using superordinate and prototypical cues should be done and controlled studies undertaken to determine intervention efficacy. Thus, effective stimulation may slow down the functional deterioration and improve the quality of life for the patient with DAT.

The study of this disease also provides a window, however cloudy, on the neurolinguistic organization of the brain and the relationship of language and thought. While we are reminded of Hughlings Jackson's admonition concerning the fallacy of confusing the location of brain damage related to a language dysfunction with the locating of that function in the normal state, it remains that we must continue to probe and study all pathologies which affect language or any other cognitive or mental ability in our search for and understanding of brain-behavior relationships.

REFERENCES

Albert, M. L. (1981). Changes in language with aging. *Seminars in Neurology, 1*:43-46.

Appell, J., Ketesz, A. & Fisman, M. (1982). A study of language functioning in Alzheimer's patients. *Brain and Language, 17*:73-91.

Bayles, K. A. (1982) The use of language tasks in identifying etiologically different dementias. Paper presented at the meeting of the International Neurological Society, Pittsburgh, Pa.

Bayles, K. A. (1985). Communication in dementia. In Ulatowska, H. K. (Ed.), *The Aging Brain: Communication in the Elderly* (pp. 157-173). Boston: College Hill Press, Inc.

Bayles, K. A. & Boon, D. R. (1982). The potential of language tasks for identifying senile dementia. *Journal of Speech and Hearing Disorders, 47*:210-217.

Bayles, K. A., Tomoeda, C. K., Kasnizk, A., Stern, L. & Eagans, K. (1985). Verbal preseveration of dementia patients. *Brain and Language, 25*:102-116.

Bayles, K. A., Tomoeda, C. K., Obler, L. K., Albert, M. L., Wertz, R. T., Au, R. & Gonzalez-Rothi, L. J. (November, 1988). Are the communication disorders of dementia appropriately characterized as aphasia? Walsh, P. (chair). Short course presented at the annual convention of the American Speech-Language and Hearing Association, Boston, Mass.

Cummings, J. L. (1985). Dementia: neuropathological correlates of intellectual deterioration in the elderly. In Ulatowska, H. K. (Ed.), *The Aging Brain: Communication in the Elderly.* Boston: College Hill Press, Inc., pp. 53-68.

Davis, G. A. (1983). *A Survey of Adult Aphasia.* Englewood Cliffs, N.J., Prentice-Hall, Inc., pp. 136-138.

Davis, G. A., & Baggs, T. W. (1985). Rehabilitation of speech and language disorders. In Jacobs-Condit, Linda (ed.), *The Gerontology and Communication Disorders.* Rockville, MD: The American Speech-Language-Hearing Association, pp. 185-243.

Gewerth, L., Shindler, A. & Hier, D. (1984). Altered patterns of word associations in dementia and aphasia. *Brain and Language, 25*:102-116.

Golper, L. A. C. & Binder, L. M. (1981). Communicative behavior in aging and dementia. In Darby, J. Jr. (Ed.), *Speech Evaluation in Medicare.* New York: Grune & Stratton, pp. 225-254.

Hier, D., Hagenlocker, K. & Shindler, A. (1985). Language disintegration in dementia: effects of etiology and severity. *Brain and Language, 25*:117-133.

Martin, A. D. (1981). Therapy with the jargon aphasic. In Brown, J. W. (Ed.), *Jargonaphasia.* New York: Academic Press, Inc.

Naeser, M. A., Helm-Estabrooks, N., Haes, G., Auerbach, S. & Scrinivasan, M. (1987). Relationship between lesion extent in Wernicke's area on computed tomographic scan and predicting recovery of comprehension in Wernicke's aphasia. *Archives of Neurology, 44*, 73-82.

Naeser, M. A., Mazurski, P., Goodglass, H., Peraino, M., Laughlin, S. & Leaper, W. (1987). Auditory syntactic comprehension in nine aphasia groups (with CT scans) and children: Differences in degree but not order of difficulty observed. *Cortex, 23*:359-380.

Nation, J. E. & Aram, D. M. (1977). *Diagnosis of Speech and Language Disorders.* St. Louis: C.V. Mosby Co.

Nebes, R. D. (1985). Preservation of semantic structure in dementia. In Ulatowska, H. K. (Ed.), *The Aging Brain: Communication in the Elderly*. Boston: College Hill Press, pp. 109-122.

Nicholas, M., Obler, L. K., Albert, M. L. & Helm-Estabrooks, N. (1985). Empty speech in Alzheimer's disease and fluent aphasia. *Journal of Speech and Hearing Research, 28*:405-410.

Obler, L., Albert, M. & Helm-Estabrooks, N. (1982). Noninformative speech in Alzheimer's dementia and in Wernicke's aphasia. Paper Presented at the Academy of Aphasia.

Overman, C. A., Geoffrey, V. C. (1987). Alzheimer's Disease and other dementias. In Mueller, H. G. & Geoffrey, V. C. (Eds.), *Communication Disorders in Aging: Assessment and Management*. Washington: Gallaudet Press, pp. 271-297.

Witte, K. (April, 1989). Dealing with dementia in long term care. Paper presented at the New York State Speech Language and Hearing Association Annual Conference, Kiamesha Lake, N.Y.

CHANGES IN MEMORY PROCESSES OF DEMENTIA PATIENTS

Taher Zandi, Ph.D.

Associate Professor of Psychology
State University of New York
Plattsburgh, N.Y.

In the past decade psychologists, psychiatrists and neurologists have become significantly interested in studying memory processes of older adults and in particular, memory processes of dementia patients. Scientists' attention to the memory processes comes at a time when clinical gerontologists are confronted with an overwhelming number of "clients" who are concerned about their memory failure. Also, the differential diagnosis of dementia disorders is still unresolved. The picture is even more bleak since the chief characteristic of the disorder; namely, changes in memory processes, are not well understood and its properties are not known.

The major objectives of this chapter are: 1) to identify the basic ingredients of memory that are important in identification of dementia and in particular, the dementia of Alzheimer's type; 2) to distinguish between normal and abnormal aging memory changes of dementia patients; and 3) to evaluate the commonly used memory measures in light of the basic structure of memory processes.

BASIC INGREDIENTS OF MEMORY PROCESSES

To remember something three things are necessary: A) we must encode the information, then we must process the information in the form of a sensory stimuli and translate it into conceptually understandable codes (this concept may vary depending on previously acquired knowledge, age, education, etc.) Piaget (1964); B) we must store the encoded information so it becomes part of our memory schemata and can be easily accessed; and C) we must be able to retrieve the information. The retrieval process requires a memory search that is usually prompted by contextual cues or an internally motivated search (Smith, 1985). Talland (1968) more closely examined the search and retrieval processes of the memory system. In his experiments on sensory-motor performance in relation to age and Parkinsonism, he found that during the retrieval phase the person has to hold the data regarding the to-be-found information in a more accessible storage while the search is continuing for the target information. This series of investigations led him to the discovery of a memory process that he later refers to as **working memory**. More specifically, the working memory is a miniature of encoding, storage, and retrieval processes which may fade away as time goes by. The longer the information remains in working memory the greater chance of a more permanent registration. Registration processes, therefore, at least from this perspective, are time dependent.

Furthermore, the registration processes are influenced by the type of and relevancy of the to-be-remembered information. In the classical studies of Bartlett

New Directions in Understanding Dementia and Alzheimer's Disease,
Edited by T. Zandi and R. J. Ham, Plenum Press, New York, 1990

(1932) we learned that the subjects who had no background about the "Indian folks" made more error of intrusions and less accurate recall, and that their correct recall improved and had less intrusion error as the number of trials increased. The more we know about the subject the easier processing it becomes.

Memory is schematic. It is a whole and its parts are connected to one another (Anderson, 1983). The activation of pieces of information leads to activation and remembering of other related information. The closer the relationship between the activated pieces of memory the better they are remembered. The information away from the center of this activation remains less manipulated and remains unrecalled (Zandi, 1986). For example, in a stimulant such as (Joe got hit by a small orange car as he was crossing at the left side of the road), when the subject is presented with a cue such as "accident" the extent of activation may be stronger within the main frame of this statement (i.e., Joe got hit by a car). The extent of activation within the non-main frame information such as the color of the car and left side of the road is less activated, therefore, they have less chance of being retrieved (Kintsch, 1974; Locque & Robin, 1986; Zandi et al., 1985).

The storage stage of memory processes has traditionally been referred to as the interval of time between encoding at input and retrieval at output when the information is held in memory (Smith, 1980). This stage is usually subject to memory failure as the result of intervening events that somehow interfere with the maintenance of the storage of the to-be-remembered information. The interferences may happen proactively, i.e., interferences from previously learned material, or retroactive, i.e., interferences from tasks interpolated between experiments (Craik, 1968, Fozard & Waugh, 1969). The multi-information-processing model approaches to learning and memory suggest two intriguing concepts describing the storage and processes of the memory system. These are storage structures and the central operation (Atkinson & Shiffrin, 1968). The multi-store approach of the central operation suggests the stored material is moved from one storage structure to another and in each transformation there are qualitative changes as well as quantitative changes in the content of to-be-remembered material. More specifically, the restructuring of the learned material is enhanced by more recently learned information. In this restructuring, the greater the overlap between the existing structures and the recently learned information the better the refining of the learned material.

A more recent interpretation of this model suggests that encoding processes are guided and in part depend on factors such as the organization of the stored material. The operational definition of this model requires a simultaneous process where the to-be-encoded information is rearranged according to the previous structure of the stored information, thus enhancing the future accessibility to this information (Tulving, 1974, Smith, 1985).

AGE RELATED CHANGES VS ABNORMAL MEMORY

In discussing the age related changes in memory processes in comparison to memory pathology, one faces a twofold task. First, the nature of age related changes has to be determined then one has to compare the "normal" changes in memory processes that are age related with the "abnormal" aspects of a memory disorder. The latter is even more complicated since "abnormal" memory is yet to be defined in the scientific community of memory researchers.

The age related changes in memory have been attributed to encoding, storage and retrieval aspects of the memory process.

In a typical example of depth of processing paradigms, subjects in different groups are told to use different strategies during or before processing of the to-be-remembered information. The commonly imposed strategies are 1) use of organization scheme--finding relationships between the to-be-encoded information as they are being processed; 2) use of elaboration techniques--a non-conceptual concentration on the to-be-encoded information; and 3) use of imagery system--use of a nonverbal system in conjunction with the verbal system. Elderly show greater organizational difficulties. As the number of trials increase, younger subjects (average age of 25) show a greater advantage compared to the older subjects [average age of 70] (Smith, 1985). Hultsch (1971) manipulated the experimental conditions of the tasks in an attempt to control the degree of organizational processing during encoding. Hultsch asked half of the subjects to sort words into two-to-seven categories prior to recall (organization condition). The rest of the subjects were control groups. In sorting, Hultsch found no differences between the age groups. However, differences were found in the delayed recall performance. Findings from this study suggests that the age differences in organization might be only a "retrieval deficit" (Reese, 1976). The older adults, in the retrieval stage, failed to organize the to-be-recalled information, but not because they did not have the initial organization of the younger adults at the encoding stage. Other research showed that proper instruction (i.e., mnenomics) can compensate for the organization deficit of older adults. (Craik & Tulving, 1975).

RETRIEVAL AND STORAGE, ROLE OF AGING

Before the late 1960s, the most popular hypothesis explaining the memory deficit of older adults was storage deficit, which suggested that there are interferences in storage processes that causes displacement of information and loss. The recent popular view replacing the storage hypothesis is the retrieval hypothesis. The retrieval hypothesis attributes the memory deficit to failure of accessibility at the time of the test--cued dependent forgetting rather than trace dependent forgetting (Tulving, 1974). Retrieval processes are examined through the use of a variety of comparisons. Among the most commonly used methods are: comparison of recall and recognition and comparison of free recall and cued recall. Students of memory have almost unanimously suggested that the age differences found in free recall was not found when recognition tests were used. Tulving, in a recent work (1983), recognized that retrieval processes are influenced by **the source of information, type of information,** and **the nature of information**. Although Tulving has used other terms describing the variables that influence retrieval processes (pp 35), I find it more suitable for the applied memory scholars to learn this by the names I selected.

The **source of information** has to do with the modality that the subject has to use in order to register the information and later on retrieve it. The sensory system is always involved in the processing system and we all agree, is the very first stage of processing. Information that is retained at this level will certainly have a more difficult retrievability than the information that is comprehended. As Tulving suggested, "A mere sensation of some perceptual inhomogeneity in an otherwise homogeneous perceptual field is sufficient as the source of information" (p. 36). Whereas comprehension of this information needs registration and understanding of the meaning of this information, the mere sensation, although symbolic (e.g., pair of words), does not represent meaning and in most cases older adults perform poorly on this type of task. Yet, they perform equally well on tasks that are semantic related in comparison to younger adults.

Warrington and Weiskrantz (1982) observed patterns of results in their amnesia patients similar to the retrieval ability of older adults as described above. Warrington and Weiskrantz (1982) found that their patients who had great difficulty remembering recently studied material had less difficulty making use of the information that they had acquired in the course of studying the material. For example, a patient may not be able to identify a specific test word in the retrieval condition, but may show evidence of having acquired that word in other tasks such as problem solving. He/she can produce the name of the word in response to a preceptual fragment of it more readily than he/she could have done in the absence of the learning experience. More specifically, when the patient had a chance to produce a meaning with the word, then comprehension took place and retrieval was enhanced.

The **type of information** is referred to as the typicality of information. The more atypical (uncommon) the format of the to-be-remembered information, the more difficult its retrieval becomes. If the to-be-remembered information is to be included by associating it to our personal experiences at the encoding level (episodic memory), then one will have a hard time doing so if the task involves remembering the name of three semantically unrelated objects and recalling them later on (e.g., this is part of the memory assessment of the Mini-Mental State Examination of Folstein et al., 1975).

The **nature of the information** has to do with how factual the information is. Factual information is more easily remembered in comparison to non-factual information. The sky is blue (even on a cloudy day) is factual information. However, who the president of the United States is may not be a factual statement or as factual as the previous one: a) the encoder did not follow the presidential debates; b) if the election had just taken place; and/or c) if the election process is not part of one's personal experience (Tulving, 1983).

In general, the commonly found age differences that are reported in memory and aging literature decline when the elderly subject is presented with a meaningful task that is not temporal in nature and is commonly available (Tulving, 1974; Walsh, 1980; Hultsch, 1982).

In conclusion, the scholars of memory among dementia patients have to consider the fact that failure in memory tasks is not the inevitable result of a person's progress of the disease. It is quite possible that it is the assessment process that may hamper retrieval processing.

ABNORMAL ENCODING

As it is noted, tests of encoding processes are extremely difficult tasks because one has to evaluate encoding processes based on what a subject has retrieved. The encoding difficulty in clinical settings are evaluated in light of two things: A) having efficient brain mechanisms for comparing representation in memory systems that require simultaneous storage, organization and restructuring; and B) having the appropriate modes of retrieval in association with working memory (Hunt, 1986). Amnesia patients have shown proper memory of recent information and rather intact remote memory. Squire (1981), indicated that retrograde amnesia can temporarily fade and that recently formed information is more vulnerable to dissipation. In a study of antrograde amnesia subjects who had received as many as five bilateral ECT treatments only a few hours before being presented with a short story, had the same recall as the control group on the immediate recall. However, their recall of the same story was significantly less accurate one day after the encoding (Squire et al., 1984).

These findings suggest that the encoding of the to-be-remembered information takes place, since the antrograde amnesia patients have no difficulty recalling the recent information shortly after presentation and that the retrograde patient experiences interferences that must be caused by more recently encoded information. The same condition is seen in the Korsakoff patients even though the neuropathology· of antrograde amnesia and Korsakoff's are quite different (Miller, 1977; Butters, Martone, White, Granholm & Wolf, 1985).

Dementia of Alzheimer's type patients (AD) shows no specific difficulty in encoding information. Like Korsakoff patients, their difficulty seems to be more related to restructuring of stored information and retrieval. Butters et al. (1984 & 1987) evaluated the beneficial effects that verbal mediation and labeling might have on amnesic, Huntington's, Korsakoff's, and demented patients' ability to remember pictorial material. Two conditions were employed: no story condition followed by a story condition. Results of this study suggested that the four patient groups were aided, however differently, by the presentation of verbal mediation. The Alzheimer's Disease and Korsakoff patients equally performed worse than the Huntington group. However, when the verbal mediator was not included Alzheimer's Disease patients performed on the same level as the other groups on forced picture recognition. Zandi & Woods (1988) studied the Alzheimer's Disease patient in comparison to the normal elderly subjects matched by age and education. Two conditions were employed in encoding and two conditions were employed in retrieval. Subjects in group one were presented with a sentence describing an event and group two subjects heard the sentence and looked at a picture depicting the verbal information in a test of free recall. Control subjects outperformed the Alzheimer's Disease patients in both conditions. However, the gap concerning the performance of Alzheimer's patients and the control group became much smaller when they were given a three-choice recognition questionnaire. The differences grew even smaller in the sentence/picture condition for the control and Alzheimer's Disease patients.

What seems to be the abnormality in memory processes of the Alzheimer's Disease seems not to be related to their encoding. Their memory performance improving in the recognition condition suggested that information was processed (encoded) yet not easily accessed and retrieved. Furthermore, at least for the immediate recall condition, these findings suggest that information is stored and it is the retrieval processes that should be a target of investigation.

ABNORMAL STORAGE AND RETRIEVING

The majority of evidence concerning the storage difficulty of normal older adults is concentrated on the organization factor of storage. Generally, the older adults' difficulty seems to be of the production defray or inefficiency variety. That is, older adults do not spontaneously use organizational strategies as extensively as younger adults (Eysenk, 1974; Huylicka & Grossman, 1967; Craik et al., 1975; Walsh, 1980). However, when various organizational strategies are built into the situations, the performance of older adults improves. Older adults' memory strategies may differ from younger adults. Then, once they learn the new strategies they compensate for their lack in memory performance (Zandi, 1982).

Hart, Smith & Swash (1985) examined recognition memory for several types of stimuli material in patients clinically diagnosed as having early Alzheimer's Disease and in the matched control group. Alzheimer's Disease patients showed significant deficits in recognition (naming) of verbal and abstract geometric shapes than the control groups. However, the Alzheimer's patient's memory of faces was the same as that of the control groups. In other words, the observed deficits depended also on the nature of the stimuli. Does this suggest that some part of the memory structure is damaged and causes failure in naming abstract items? Or, is this a selective task and the dementia patient prefers to name pictures? Is the quantity of recall (how many items) an accurate criteria of testing of memory or does one have to look at the qualitative aspect of memory as well?

Although examining how much information is recalled clearly allows for statements regarding memory dysfunctions, this does not provide us with what actually occurs in terms of the clinical perspective of memory processes (Schultz, Schmitt, Logue, Rubin, 1986). Rubin, Olson, Richter and Butters (1981) studied memory for prose in both Korsakoff and schizophrenic patients. They found the memory impairment is more severe if measured quantitatively rather than qualitatively. Zandi et al. (1987) demonstrated that normal elderly recall of the most important units of a prose is the same or better than the younger adults. However, younger adults' recall of trivial information (detail) was greater than normal elderly. In a 1986 study Schultz et al. developed a test composed of quantitative and qualitative aspects of the logical memory and visual reproduction tests of the Wechsler Memory Scale (WMS) and Russell's revision of WMS (RWMS). Six groups of subjects were used. These were Alzheimer's patients, multi-infarct patients, head injury patients, metabolic disorder patients, affective disorder patients and normal control groups.

Using the quantitative measures of prose recall, Alzheimer's Disease patients should have the poorest performance and the metabolic and the affective groups, as well as, head injury group should perform the same. Using the qualitative measure (major ideas of the prose) the Alzheimer's Disease patients recalled the same major ideas. This finding is of particular significance because it raises the question of the functional capacity of the semantic vs episodic memory. Is the Alzheimer's Disease patient's recall guided by their semantic (knowledge) memory? Findings as such, suggest expansion of a selective retrieval process that is guided by a form of organizational strategy.

Martin & Fedio (1983), Wington, Grafmen, Boutelle, Keye & Martin (1983) found that both amnesia and demented patients are impaired in the acquisition and recall of material associated with particular temporal and/or spatial context (i.e., episodic memory). Furthermore, demented patients are seriously impaired in remembering semantic information (general knowledge). (See Tulving, 1983 for a detailed description of espisodic vs semantic memory.)

Korsakoff patients and demented patients performed poorly on tasks that required acquisitions of new knowledge (Butters, Granholm, Salmon & Grant, 1987). The Korsakoff patients, however, perform better on semantic tasks than on episodic tasks. When testing the semantic memory, proactive interferences were greater among Korsakoff patients (Butter & Germak, 1980) and they were more prone to errors of intrusion than demented patients.

In a recent study Butters et al. (1987), compared the memory performance of Alzheimer's patients on semantic tasks (letter and category fluency) and episodic tasks (recall of passage) with Korsakoff and Huntington patients matched for overall severity of dementia. In this study, two control groups of normal young and older adults were utilized. The Korsakoff and Alzheimer's patients produced more intrusion errors than other groups of subjects. In general, the Alzheimer's patients showed difficulty with both episodic and semantic memory tasks which differentiated from the Korsakoff and Huntington patients. Alzheimer's patients showed the greatest vulnerability to interferences when they were asked to produce names of animals while seeing their picture (visual fluency). Alzheimer's patients experienced a great deal of difficulty, but they encountered very few problems with letter fluency tasks. Butters et al. (1987), suggested that the Alzheimer's patients' deficits in semantic memory were most apparent when they had to search for examples of an abstract category (i.e., animals) or when they were presented with a subordinate from that category.

Butters' et al. (1987) study like Ober et al. (1986) and Martin & Fedio (1983) suggest that Alzheimer's patients' language problems involve a reduction in the number of examples forming an abstract category. Therefore, for these patients, scores on the category fluency task should be a highly sensitive measure of deficiency in semantic memory.

It is rather reasonable to conclude that the major failure in dementia patients occurs during the search processes for retrieval of information. Nevertheless, the working memory (memory that holds the data in mind in order to give action until a task is completed) is not fully operational, thus creating no organization and no completion in the searching processes.

MEMORY DYSFUNCTION EVALUATION TASKS

Memory assessments are basically designed to evaluate:

a) the **neuropathological aspect** of memory. These assessment instruments are geared toward locating the source of memory dysfunction as this relates to the central nervous system (CNS). These assessment instruments are quite useful in teaching us about the existence of tumors, vascular conditions (e.g., aneurism of the anterior communicating artery) (Goldberg, Bilder, 1986).

b) **experimental descriptive models** of memory failure (Hunt, 1986). The assessment instruments from this perspective are put together based on a theoretical model that describes cognitive deficits. The most prominent descriptive model in recent years has been **the productive activation model** that identifies the specific operating system of working memory and/or long-term memory (Newell, 1973)

c) the clinical **observational approach** in which the clinician compares observed information (data fitting) with the available neuropathological and psychological theories. See Table 1 for the listing of examples from each category.

The scholars of memory disorders are aware that the primary purposes of the assessment instruments are: 1) to contribute to the medical diagnosis of the syndrome (e.g., Korsokoff vs Alzheimer's Disease); 2) to contribute to the screening for localization of brain lesions; and 3) to contribute to the treatment and management of individuals with memory problems (Poon, Garland, Eisdorf, Crook, Thompson, Kaszniak & Davis, 1986).

The neuropsychological assessment instruments are used mainly to assess patients with possible amnesia disorders and are used for short-term and long-term follow-up of amnesia patients. In general, they appropriately differentiate between the memory dysfunctions that are tumor related and prolonged dementia. Specifically, these instruments are quite useful in differentiating between the symptoms that have a neurological origin particularly given the similarity that exists between the memory dysfunction of amnesia and dementia syndrome.

The clinical pictures of both these disorders may include disorientation to time and space, anomia, short sentences, apraxia, profound loss of recent memory, profound retrograded and amnesia (e.g., recognized distant cousin whom he knew for the past 60 years but yet does not recognize his children). At first glance, the majority of the memory difficulties listed above signal the possibility of a moderate to severe dementia. At the same time this could be the memory profile of a 40-year-old person with an open skull fracture in the right parieto-occipital and temporal area who has just come out of a coma with his computerized tomography revealing ventricle enlargement. Administration of neuropsychological tests will allow the clinician to keep track of the patient's prognosis on both the short-term and long-term basis and distinguishes amnesia from dementia based on the extent of patient recovery. The majority of assessment instruments that are constructed based on **experimental descriptive models** of memory dysfunction, assume that thoughts are formed based on internal representation and external environmental stimulation of the processing system (Newell, 1973).

According to this model, the function of the working memory is to store recent information and exchange that with the long-term memory which stores remote information. The assessment instruments are, therefore, designed to evaluate the functional capacity of both the working memory and the long-term storage process. The measures of the working memory structure and its dysfunctions are both verbal and nonverbal. The size of the working memory is generally measured by a memory span test (e.g., Wechsler Memory Scale). Furthermore, other measures such as remembering the names of a few objects, counting from 100 backwards by sevens or spelling "world" backwards all require that the subject be able to hold certain information in the working memory while changing its product (e.g., Mini-Mental State Examination, Folstein, 1975). See Table 1 for a listing of other memory questionnaires constructed based on this model.

The major criticism of the tests of working memory is that people are only asked to keep information in their working memory in isolation. The working memory defined previously in this chapter operates based on the holistic representations of to-be-remembered information. Therefore, as Hunt (1986) **and** Clenzer, Fischer, & Dorfman (1984) suggested, the working memory capacity should be tested in contexts that are important in a person's life and that are related to the person's past memory processing activities. The majority of linguistic tests are advantageous in testing impairment of working memory since a majority of our processing activities are in the form of language comprehension. Specifically, as we discussed before, tests of propositions that are semantically or episodically relevant to one's primary system seem to be of greater value. For example, Glanz et al. (1984) demonstrated that normal readers retain in their working memories two sentences verbatim. A test of

TABLE 1

Examples of Memory Dysfunction Evaluation Tasks
and Their Corresponding Memory Models

Instrument	Measure	Tasks Involved	Model	Purpose	Author
Wechsler Memory	Verbal Memory	Short-term/long term Encoding, Retrieval Recent Memory	Neuropsychology	Differential Diagnosis, Brain Leison	Wechsler (1945)
Buschke Selective Reminiscing	Verbal Memory	Recent Memory	Neuropsychology	Familiar Remote Memory	Buschke (1973)
Boston Retrograde Amnesia	Episodic Memory Verbal & Non-Verbal	Recognition Famous Faces	Neuropsychology	Differential Diagnosis	Albert, et al. (1979)
Wisconsin Card Sorting, Trail Marking, Word Fluency	Executive System Verbal	Recall & Recognition	Clinical Observation & Neuro-Psychology	Language Production	Wisconson
Cambridge Examination for Mental Disorders of the Elderly	Global	Neuro-Behavioral Clinical	Neuropsychology Clinical Observation	Dementia Diagnosis Differential Diagnosis	Roth et al. (1988)
Mirror-Reading Task	Verbal Reading	Semantic, Factual Information	Neuropsychology	Differential Diagnosis of Amnesia & Dementia	Cohen, al. (1980)
Metamemory Questionnaire	Verbal, Recall	Frequency of forgetfulness, Use of Memories	Experimental	Self-Awareness, General Diagnosis	Zelinski et al. (1980)
Short Portable Mental Status Questionnaire	Verbal & Non-Verbal	General Mental Status	Clinical	Psychiatric Diagnosis of Organic Brain Syndrome	Pfeiffer (1975)
Blessed-Tomlinson-Roth Information Concentration Test	Verbal & Non-Verbal	General Mental Status	Neurological	Cognitive Deterioration Related to Cerebral Gray Matter	Blessed, (1968)
Mini-Mental Status Exam	Verbal & Non-Verbal	Memory/Orientation Attention, Paraxis Language	Neuropsychology	Diagnosis Dementia & Affective Disorder	Folstein (1975)
Mental Status Questionnaire	Verbal & Non-Verbal	General Mental Status	Neuropsychology	General Diagnosis of Organic Brain Syndrome	Khan (1970)
Inventory of Memory Experience	Verbal & Non-Verbal	Frequency of Forgetting & Remembering	Experimental & Descriptive Model	Identification of Dementia	Herrmann et al. (1978)
Wadsworth Memory Questionnaire	Verbal	General Mental Status	Experimental Descriptive Model	Identification of Demenita	Goldberg (1981)

ability to retain information in working memory while reading could be of greater value (Daneman, 1983).

The measurements that are based on data fitting and are done in **clinical observation** are in general concerned with three things:

1) The characteristics of individuals--their processing abilities, education, skills, knowledge and occupation. Access to this information will allow the observer to create a baseline measure prior to integrating the recently collected data. The greater the discrepancy between the memory complaints reported by the patient and the caregiver, the more likely there is a memory difficulty (Poon et al., 1986).

2) The sensitivity of the test and the dependent measure that determines the degree of dysfunction (e.g., recall vs recognition or test of encoding vs retrieval), as well as the subjects' degree of familiarity with memory tasks.

One has to remember that all the assumptions concerning cohort differences that can make a memory task biased for a particular age group can also be biased for a group of memory impaired elderly subjects. More specifically, because the subject's memory is impaired the task criteria and its biases are not to be ignored. See Table 1 for a listing of a few examples of measurements that have been constructed based on the clinical observation assumption.

The issue of memory assessment of dementia patients and other memory impaired people is far from being resolved. The inconsistencies that exist within literature and among the students of memory in my judgment is due to the following sources:

1. There have been no serious dialogues and exchanges between the students of memory and clinicians who inadvertently discover memory impairments in their patients. The latter group assesses the incidental findings with instruments that may be non-specific to memory loss.

2. A majority of memory assessment instruments ignore the individual differences that exist among memory impaired individuals. The memory symptoms dictate the selection of memory assessment instruments as well as their construction, regardless of the etiological, neuropsychological and pathological description of the memory impairment.

3. Examination of memory processes includes tasks that are not memory related. Therefore, what is being measured is something different than memory. For example, the patient's lack of comprehension of test instructions may be interpreted as memory failure.

4. The test selection procedure does not include hypothesis testing criteria. The investigator needs to know whether he/she is testing problems in acquisition of new information, or if memory searching difficulties are being investigated. The existence of the hypothesis testing situation will give direction to the construction of these instruments.

5. Finally, the clinicians and neuropathologists need to work more closely with each other in order to match the behavioral memory symptoms with what the pathologists find on the autopsy table or obtain through other scientific inquires.

Cooperation among the scholars of memory should be increased as we are preparing to confront the memory dysfunctions of elderly in the next decade and into the next century. The prevalence of all forms of memory impairment, in particular dementia, will increase as the number of older adults, specifically, those over age 85 quadruples by the second decade of the next century.

REFERENCES

Albert, M. S., Butters, N., Levin, J. (1979). Temporal ingredients in the retrograde amnesia of patients with alcoholic Korsakoff's disease. *Archives of Neurology, 36*:211-216.

Anderson, J. R. (1983). A spreading activation theory of memory. *Journal of Verbal Learning and Verbal Behavior, 22:261-295.*

Atkinson, R. E., & Shiffrin, R. M. (1968). Human memory: Proposed system and its control process. In K. W. Spence & J. T. Spence (eds). The *Psychology of Learning and Motivation.* (Vol. 2). New York: Academic Press.

Bartlett, F. C. (1932). *Remembering.* Cambridge: Cambridge University Press.

Blessed, G., Tomlinson, B. E., Roth, M. (1968). The association between quantitative measures of dementia and senile changes in the cerebral gray matter of elderly subjects. *British Journal of Psychiatry, 114*:797-811.

Buschka, H. (1973). Selective reminding for analysis of memory and learning. *Journal of Verbal Learning & Verbal Bahavior.*

Butters, N. & Germak, L. S. (1980). *Alcoholic Korsakoff's Syndrome: An Information Processing Approach to Amnesia.*

Butters, N. (1980). Potential contribution of neuropsychology to our understanding of the memory capacity of the elderly. In L. W. Poon, J. L. Fozard, L. S. Cermak, D. Arenberg & L. W. Thompson (Eds), New *Directions in Memory and Aging.* Lawrence Erlbaum Associates, Hillsdale, N.J.

Butters, N., Miliotis, P., Albert, M. S. & Sax, D. (1984). Memory assessment: evidence of heterogenity of amnesiac symptoms. In G. Golstein, *Advances in Clinical Neuropsychology.* (Vol. 1). New York: Plenum Press.

Butters, N., Martone, M., White, B., Granholm, E., Wolf, J. (1985). Clinical validations: Comparisons of demented and amnesiac patients. In L. W. Poon (ed), *Clinical Memory Assessment of Older Adults.* APA. Washington, D.C.

Butters, N., Granholm, E., Salmon, P. D. & Grant, I. (1987). Episodic and semantic memory: A comparison of amnesiac and demented patients. *Journal of Clinical and Experimental Neuropsychology, 9* (5):479-497.

Clenzer, M., Fischer, B. & Dorfman, D. (1984). Short-term storage in reading. *Journal of Verbal Learning and Verbal Behavior, 23*(4):467-486.

Craik, F. I. M. (1968). Short term memory and the aging process. In G. A. Tallard (ed), *Human Aging and Behavior.* New York Academic Press.

Craik, F. I. M. & Tulving, E. (1975). Depth of processing and the retention of words in episodic memory. *Journal of Experimental Psychology General, 104*:268-294.

Daneman, M. (1983). The measurement of reading comprehension. How to trade construct validity for predictive power. *Intelligence, 6*:331-345.

Eysenk, M. W. (1974). Age differences in incidental learning. *Developmental psychology, 10*:936-941.

Eysenk, M. W. (1977). *Human Memory: Theory, Research, and Individual Differences.* Oxford: Pergamon Press.

Folstein, M. F., Folstein, S. E., McHugh, P. R. (1975). "Mini-Mental State" A practical method for grading the cognitive state of patients for the clinician. *Journal of Psychiatry Research, 12*:189-198.

Fozard, J. L. & Waugh, N. C. (1969). Proactive inhibition of prompted items. *Psychonomic Science, 17*:67-68.

Goldberg, E. & Bilder, M. R. (1986). Neuropsychological perspective: retrograde amnesia and executive deficits. In L. W. Poon (ed), *Clinical Memory Assessment of Older Adults.* American Psychological Association, Washington, D.C.

Goldberg, Z., Syndulko, K. K., Lemon, J., Montan, B., Ulmer, R. & Tourtellotte, W. W. (1981). Everyday memory problems in older adults. Paperpresented at the meeting of American Psychological Association, Los Angeles.

Herrmann, D. J. & Neisser, O. (1978). An inventory of everyday memory experiences. In M. M. Arunberg, P. E. Morris & R. N. Sykes (eds) *Practical Aspects of Memory*. New York: Academic Press.

Hulicka, E. M. & Grossman, J. L. (1967). Age-group comparisons for the use of mediators in paired associate learning. *Journal of Gerontology, 22*: 46-51.

Hultsch, D. F. (1982). Adult age differences in the organization of free recall. *Developmental Psychology, 1*:673-678.

Hultsch, D. F. (1971). Adult age differences in free recall. *Developmental Psychology, 4*:338-347.

Hunt, E. (1983). On the nature of intelligence. *Science, 219*:141-146.

Hunt, E. (1986). Experimental perspectives: Theoretical memory models. In L. W. Poon (ed), *Clinical Memory Assessment of Older Adults* . American Psychological Association, Washington, D.C.

Khan, R. L., Goldfarb, A. I., Pollack, M., Peck, A. (1960). Brief objective measures for the determination of mental status in aged. *American Journal of Psychology, 117*:326-328.

Kintsch, W. (1974). The representation of meaning in memory. Hillsdale, N.J.: Lawrence Erlbaum Association.

Martin, A. & Fedio, P. (1983). Word production and comprehension in Alzheimer's disease: The breakdown of semantic knowledge. *Brain and language, 19*:124-141.

Miller, E. (1977). Abnormal aging: the psychology of senile and presenile dementia. New York: Wiley.

Newell, A. (1973). Production system: Models of control structures. In W. G. Chase (ed), *Visual Information Processing*. New York: Academic Press.

Ober, B. A., Dronkers, N. F., Koss, E., Delis, D. C. & Friedland, R. P. (1986). Retrieval from semantic memory in Alzheimer-type dementia. *Journal of Clinical and Experimental Neuropsychology, 8*:75-92.

Pfeiffer. E. (1975). A short portable mental status questionnaire for the assessment of organic brain deficits in elderly. *Journal of American Geriatric Society, 23*:433-441.

Piaget, J. (1964). *Experiments in Contradiction*. Chicago: University of Chicago Press.

Poon, L. W., Garland, B. J., Eisdorfer, C., Crook, T., Thompson, W. L., Kaszniak, N. A. & Davis, L. K. (1986). Integration of experimental and clinical preapts in memory assessment: A tribute to George Talland. In L. W. Poon (ed), *Clinical Memory Assessment of Older Adults*. Washington, D. C.: American Psychological Association.

Reese, H. W. (1976). Models of memory development. *Human Development, 19*:291-303.

Rubin, D. C., Olson, E. H., Richter, M. & Butters, N. (1981). Memory for prose in Korsakoff and schitzophrenic populations. *International Journal of Neurscience, 13*:81-85.

Schultz, A. K., Schmitt, A. F., Logue, E. P. & Rubin, D. (1986). Unit analysis of prose memory in clinical and elderly population, 2(2):77-87.

Smith, A. D. (1980). Age differences in encoding, storage, and retrieval. In L. W. Poon, J. L. Fozard, L. S. Cermak, D. Arenberg, L. W. Thompson, *Directions in Memory and Aging*. Lawrence Erlbaum Associates, Inc., N.J.

Smith, J. (1985). Acquisition of memory skill by older adults. Paper presented at the Eighth Biennial Meeting of the International Society for the Study of Behavioral Development. Tours, France.

Squire, L. R. (1981). Two forms of human amnesia: An analysis of forgetting. *The Journal of Neuro-Science, 1*:635-640.

Squire, L. R. & Butters, N. (eds) (1984). *Neuro-Psychology of Memory.* New York: Guilford.

Talland, C. A. (1968). Age and the immediate memory span. In G. A. Talland (ed), *Human Aging and Behavior.* New York Academic Press.

Tulving, E. (1974). Cue dependent forgetting. *American Scientist, 62*:74-82.

Tulving, E. (1983). *Elements of Episodic Memory.* New York: Oxford University Press..

Tulving, E. & Donaldson, W. (1972). *Organizing Memory.* New York: Academic Press.

Walsh, D. A. (1980). Age differences in learning and memory. In D. S. Woodruff & J. E. Birren (eds), *Aging: Scientific Perspective and Social Issues.* New York: D. Van Nostrand Co.

Warrington, E. K. & Weiskrantz, L. (1982). Amnesia: A disconnection of syndrome? *Neuropsychology, 20*:233-248.

Wechsler, D. A. (1945). A standardized memory scale for clinical use. *Journal of Psychology, 19*:87-95.

Wington, H., Grafmen, J., Boutette, W., Keye, W. & Martin, P. R. (1983). Forms of memory failure. *Science, 221*:380-382.

Zandi, T. (1982). Adult age differences in use of pictorial and prose information. An unpublished dissertation.

Zandi, T., Begin, A., Hassel, S., Tretter, A. & Woods, S. (1985). Behavioral anatomy of older adults' memory structure: A developmental view. Paper presented at the annual meeting of the Gerontological Society of America. New Orleans.

Zandi, T. (1986). Comprehension of prose information in late adulthood. Paper presented at the annual meeting of New York State Association of Gerontological Education.

Zandi, T., Castle, J., Meldrin, A. (1987). Encoding strategies of young and older adults: A primary study. Paper presented at the annual meeting of the Eastern Psychological Association, Washington, D.C.

Zandi, T. & Woods, S. (1988). A new look at the communicative strategies with Alzheimer's patients through the intact part of their memory. *American Journal of Alzheimer's Care and Related Disorders. Nov./Dec.,* 1988.

Zelinski, E. M., Gilewski, M. J., Thompson, L. W. (1980). Do laboratory tests relate to self-assessment of memory ability in the young and old? In L. W. Poon, J. L. Fozard, L. S. Cermak, D. Arenberg, L. W. Thompson, *New Directions in Memory and Aging.* Hillsdale, N.J.: Erlbaum.

Section III

Alzheimer's Disease:
The Other Victims

UNDERSTANDING PLACEMENT OF THE DEMENTED ELDERLY

Howard Bergman, MD, CCFP, CSPQ

Assistant Director, Division of Geriatrics
Sir Mortimer B. Davis - Jewish General Hospital
Assistant Professor, Departments of Family Medicine and
Medicine, and McGill Centre for Studies in Aging
McGill University
Montreal, Canada

Mrs. K., an 85-year-old widow living alone with the help of her daughter, has been your patient for many years. In spite of mild cognitive deficits she has managed to live at home and accomplish her activities of daily living with some input from her daughter for shopping and banking. Her daughter, your patient as well, brought her mother in because Mrs. K. has developed more than occasional urinary incontinence and seems to have more difficulty taking care of herself. The daughter would like you to sign the papers for application to place Mrs. K. in a nursing home.

WHEN TO PLACE YOUR ELDERLY PATIENT--WHY POSE THE QUESTION?

At first sight, the answer seems a bit self-evident. Mrs. K., an elderly patient, needs to be placed because she cannot live independently at home, and the daughter does not feel she is able to look after her anymore.

The elderly want to remain independent at home. They have become an important sector of the population with political and economic clout. Moving into a group home and in particular into a nursing home is seen as a serious loss.

Our conception of the evaluation and treatment of the elderly has considerably evolved. In fact, geriatrics grew out of the realization by Dr. Marjorie Warren in the 1940s in England that many elderly were admitted to long term care without proper evaluation, and that many of them would respond well to active treatment and rehabilitation.

Finally, our society is getting older and the inappropriate placement of the elderly, which went unnoticed 14 years ago, is becoming an important financial burden to the individual elderly person, his/her family and society in general (1).

HOW OLD IS OUR SOCIETY?

Table 1 indicates the aging of the American population. At the present time about 12% of the population is over 65 years old. By the year 2040, however, this figure will almost double. What is probably more significant, though, is that the number of elderly over 85 years old will increase by 400%.

New Directions in Understanding Dementia and Alzheimer's Disease,
Edited by T. Zandi and R. J. Ham, Plenum Press, New York, 1990

Although aging is not synonymous with illness, the prevalence of chronic disease and disability does increase with age and especially in the old-old, that is, the population over 75 years old (2). It is estimated that 80% of the elderly over 65 living at home suffer from one chronic illness, and that 50% suffer from two or more medical problems and often one psychiatric problem (3). Up to 15% of the elderly over 65 years old suffer from senile dementia (4), and this figure goes up from 20% to 47% for the elderly over 80 years old (5).

WHERE DO THE ELDERLY LIVE?

Over 5% of the elderly live in a nursing home (6). On the bright side, over 90% of the elderly live at home. Of this percentage, about 10% are maintained at home with significant home care services. Another 5% or 10% receive some kind of services which help them maintain their independence.

The decision to place an elderly person is a crucial one. For the elderly involved, it represents the loss of independence, of intimacy and of control over his/her environment. For the family it creates a sentiment of relief, but also guilt as well, and may stir up serious conflicts.

For society in general, our health care system is under stress. There is no doubt that the population in nursing homes is older, sicker and more disabled than 15 years ago. Fifty-seven percent of all nursing home admissions come from another health facility. Ninety-one percent of all nursing home patients are long term care patients. Of the remaining 9%, 75% will either die or be readmitted to an acute care hospital. Twenty-five percent will return home, but only 7% of those discharged home will be there two years later (7). In fact, in the best academic nursing homes, 16%-25% of the residents are admitted to acute care hospitals per year because of acute care illnesses which cannot be dealt with in the nursing home (8).

THE DECISION TO PLACE

The primary care physician is the person who plays a key role in the decision to place (9). He/she is usually the first to notice the deterioration of the patient or is the first to be approached by a family on the question of placement. A primary care physician is usually in contact with the children of the patient who may be his/her patient as well. Generally, therefore, he/she can best determine if the elderly person can be maintained safely at home or if institutionalization has become necessary.

However, in order to make that decision, the physician must know the elderly very well from a medical and social point of view and have a basic understanding of the principles of geriatric medicine. The physician must also know the community resources and be able to work with them as well as having an understanding of the institutionalization system in the area (10).

RISK FACTORS FOR INSTITUTIONALIZATION

There is no set formula on when an elderly should be institutionalized, but the decision is a difficult and complex one based on the factors which determine whether an elderly can continue to live at home. There are a certain number of risk factors for institutionalization which have been identified (11). These risk factors are the following: age over 80 years old, living alone, cognitive deficits, mobility with aids, and recent hospital admissions in E.R. visits (Table 2).

Knowing that a patient has one or more of these risk factors means that they must be followed more carefully and that further input from home care or

TABLE 1

U.S. Population 1900-2040

Year	Total pop (millions)	Percent 65-74 yrs.	Percent 75-84 yrs.	Percent > 85 yrs.	All ≥ 65 yrs. Number %	
1900	76 m	2.9%	1.0%	0.2%	3.1 m	4%
1960	179 m	6.1	2.6	0.5	16.6 m	9.2
1980	227 m	6.9	3.4	1.0	25.5 m	11.3
2000	268 m	6.6	4.6	1.9	35.0 m	13.1
2020	296 m	10.0	4.8	2.5	51.4 m	17.3
2040	308 m	9.5	8.0	4.2	66.6 m	21.6

(Source: U.S. Census Bureau Reports and Projections, Series P-25, 1982)

geriatric services may become necessary. Identification of these risk factors may permit the initiation of certain "preventive" measures: friendly visitors, or golden age activities for the isolated elderly, physiotherapy for those with difficult mobility, and closer medical follow-up for elderly with repeated emergency room visits and admissions.

Generally, the caregiver (a spouse or a child) will often continue to live with the elderly suffering from dementia until certain specific problems arise. For example, a spouse or a child may have great difficulty dealing with aggressive behaviour, wandering or day-night reversal. (Chapter 10 has discussed the difficulties associated with the caregiving for dementia patients.) Similarly, families who provide round-the-clock physical care to their demented relative, express a greater amount of difficulties and intolerance in dealing with incontinence. In the same vein, a spouse or a child will look after an elderly parent with decreased mobility until transfer from bed to chair or getting out of a chair requires significant assistance and until falling becomes a consistent problem. Appearance of a certain number of specific problems will very often lead the family or social worker to request placement (Table 3).

MATCHING RESOURCES TO NEEDS

Among other things, the decision to institutionalize a patient is based on whether the patient can continue to live at home independently. This decision will be also based on the examination of a series of factors taking into account the patient, his/her immediate environment and the personal and social support system (Table 4).

Although the basic consideration is, of course, the functional capacity of the patient, the final decision as to whether that person can remain at home will depend on the possibility of matching resources, including personal and social support

TABLE 2

Risk Factors for Institutionalization

1. Over 80 years old

2. Living alone

3. Cognitive deficits

4. Mobility with aids

5. Recent hospital admission or E.R. visit

systems, to the needs of the elderly person. At the outset, the patient's functional capacity must be analyzed in relation to his/her environment, in particular to the organization and layout of the home as well as the support system, including immediate family, friends and neighbours. Take the example of an 85-year-old man with severe Parkinson's disease who needs help to transfer, who cannot climb upstairs and who is mildly demented. If this man lives in a home with a kitchen on one story and the bedroom and bathroom on another story, the organization of the home itself could be a major obstacle to the person remaining at home. If he lives with his wife, who is strong enough to help him out of bed and chair and who can do the shopping and housework, then he will be able to continue living at home. Support systems may come from unexpected sources. An elderly gentleman who lived alone and absolutely needed Haloperidol 1 mg. daily in order to be able to function, was able to continue to live in the community because he would spend the day at the bakery where he had worked for twenty years. His former employee, with the help of a social worker, gave him the medication he needed every day.

Immediate support systems for family, friends and neighbours can accomplish a great deal, and families often go to great sacrifices to maintain their spouse or parent at home. However, these support systems may become insufficient. At some point, a decision to maintain an elderly at home may be based on the availability and quality of home care services. This, of course, varies from state to state and even from region to region. The physician and other professionals involved have to know the resources and work with them in setting up the home care services including homemaking, help with bathing, shopping, medication, surveillance, and so on.

Finally, the availability of medical services may be the deciding factor in a family decision to request placement (12). If the family knows the physician will be available in time of a crisis, that they will not be running back and forth to the emergency room, they may be able to tolerate the difficulty of caring for a sick family member. And the physician will be able to continue dealing with these difficult cases if he/she has support not only from the community services, but from the geriatric resources as well. Physicians who can get help easily through a phone call, rapid consultation, at home if necessary, and admission, would be willing to deal with these difficult cases.

WHEN DIFFICULTIES AT HOME INTENSIFY

Before Thinking Placement:

At some point, the physician or the family will realize that the elderly person and/or the family is not coping at home. This is often due to an acute exacerbation

TABLE 3

Problems Leading to Placement

1.	Dementia with:	-	aggressive behavior
		-	day-night reversal
		-	wandering
2.	Incontinence		
3.	Decreased mobility or ability to transfer secondary to:	-	CVA
		-	hip fracture
		-	arthritis

superimposed on a progressive decline. This event might be the development of incontinence, aggressive behaviour or wandering, decreased mobility or falls. At this point, it is necessary to organize a complete medical and functional evaluation to see if the problems cannot be controlled or reversed (Table 5). The physician cannot accomplish this medical and functional evaluation alone, and a multidisciplinary point of view is necessary (13). The physician has to work with the community resources to be able to analyze the patient's capacity to carry out activities of daily living, the behaviour at home, the family dynamics, mobility and any type of resources that might be necessary to respond to the changing situation. The physician will therefore have to work with the nurse, social workers and other professionals involved in the community services to carry out this multidisciplinary assessment.

The physician will try and determine if a new medical problem has developed. For example, is there a urinary tract infection causing the recent onset incontinence as well as decreased mobility? The medications will have to be reevaluated. Is the antihypertensive medication causing postural hypertension and leading to falls? New medication might be necessary. For example, if there are no new medical problems causing the day-night reversal, a small bedtime dose of Thioridazine might be necessary.

The home will have to be assessed carefully to see if any changes might help maintain the elderly patient at home. Mobility, after evaluation by a physiotherapist, might be improved with the use of certain aids such as a walker.

The social worker will need to discuss with the family the necessity for more home care resources in order to get help to a tired caregiver. Discussing a care giving strategy could help an extenuated spouse or child deal with the stress of taking care of a demented family member (14).

Resources outside of the home might be necessary. This could include a day centre to provide psychosocial activities for the patient as well as respite for the caregiver several days per week. Organizing admission to a respite bed for at least several weeks a year might be the necessary safety valve for a daughter or a spouse.

Geriatric Resources

Organizing such a multidisciplinary assessment might be too difficult in certain cases because of the nature of the problem itself or because of the difficulty in coordinating the interventions of the busy office-based physician with the professional involved in the community services. The physician can then refer the patient for a geriatric assessment (15) (Table 6).

TABLE 4

Factors to Take into Account When Considering Placement

1. Functional capacity of the patient.

2. Organization and layout of the home.

3. Support system   - family

                             - neighbours

                             - friends

4. Availability and quality of home care resources.

5. Availability of family doctor.

6. Availability of geriatric resources.

Geriatric resources vary from one community to another. If there are no resources at all in the community, most geriatric centres would at least accept phone calls from physicians to discuss a complex case. In the communities which have geriatric resources different levels of services might exist. On an outpatient basis the evaluation can be done in a geriatric assessment clinic or if the problem is principally psychiatric because of depression or paranoia, then referral to geriatric psychiatry would be more appropriate. Many geriatric departments now have a geriatric day hospital as part of their network of services. The day hospital permits a more ongoing evaluation of the patient from a multidisciplinary point of view while keeping a patient at home. Finally, admission to the hospital, whether to a geriatric or medical ward, might be necessary in order to attempt to identify and control the problems which seem to be leading to institutionalization.

Alternative Housing

The multidisciplinary assessment conducted in the community or in the hospital could lead to the decision that the elderly person can no longer live at home. This does not necessarily mean that the elderly person must be placed in a nursing home or in a long-term-care hospital. Before thinking of institutionalization as such, alternative housing should be looked at. In fact, the elderly should be relocated at just that level of care where their independence can be best maintained. This could be done as a trial with the physician or team being ready to transfer the patient if necessary. This situation has to be monitored by the physician or another member of the team.

Alternative housing includes low-cost housing for the elderly. In this situation, there are usually few services, but the elderly person may feel supported by the fact that he/she is living in an environment with other elderly and that a smaller percentage of his/her financial resources are going to housing. The next level of care would be the apartment/hotel. In this type of situation the elderly have their own apartment, but there is generally a maid service and a cafeteria where meals are served. In this situation there are usually minimal medical services such as the presence of a nurse on the premises as well as visits by a physician.

TABLE 5

When Difficulties at Home Intensify
Before Thinking Placement
ORGANIZE A COMPLETE MEDICAL AND FUNCTIONAL EVALUATION

---

1. Identify new medical problems and treat reversible or "controllable" disorders.

2. Medication to be discontinued, adjusted.

3. Physiotherapy, occupational therapy, physical side for mobility, transfers, etc.

4. More home care resources; strategy for caregiver.

5. Other resources   - rehab
                         - day centre
                         - respite beds, etc.

In both the low-cost housing and the apartment/hotel, the elderly must be able to maintain a good functional capacity both from a physical and cognitive point of view. When more help is needed with activity of daily living, then a group home or foster home would probably be more appropriate. Elderly patients in the foster homes generally must have socially acceptable behaviour, be continent and be able to walk from their room to the common dining room.

When Placement is the Only Alternative

Once a decision has been reached that the elderly person can no longer remain at home, the multidisciplinary assessment will have served to establish the necessary level of care. If the elderly person cannot live in an alternative form of housing in the community, institutionalization in a nursing home is necessary. This decision is often taken when the patient is in the hospital. The admission system and the admission criteria vary from state to state and even from region to region within states, but the principle of institutionalization at the appropriate level remains the same. In those regions where a centralized admissions system exists, the multidisciplinary system will permit the analysis of the application. Where these systems do not exist, the multidisciplinary assessment will give the physician and the social worker the necessary information in order to advise the family on where to apply for institutionalization.

The Pre-Admission

Admission to the nursing home could take some time. The elderly person will generally remain at home during this period unless the situation is so acute that admission to acute care hospital is necessary awaiting placement. The pre-admission period is a crucial one. The assessment will enable the physician to stabilize the patient and provide the necessary services awaiting placement.

Elderly people are often admitted to nursing homes without the proper workup, with incorrect diagnosis and with inappropriate medications (16). Patients are often admitted without the necessary medical information.

It is quite clear that the elderly admitted to long-term care institutions today are sicker, older and frailer than those institutionalized in the past. Coupled with the fact that admission itself constitutes a crisis, it is not uncommon for an elderly patient soon after admission to long-term institution to become medically unstable (17) .

TABLE 6

Geriatric Resources

1. Outpatient geriatric assessment clinic (including domiciliary assessment)

2. Geriatric psychiatry clinic (including domiciliary assessment)

3. Geriatric day hospital

4. Geriatric assessment in-patient unit or geriatric rehab unit

It is therefore essential that the pre-admission period, while the patient is awaiting placement, provide the opportunity for the physician to stabilize the medical situation of the patient, to be sure that the diagnosis and problem list are as precise as possible and that the patient is receiving the appropriate medication.

As well, the complete file should be put together with the appropriate medical history and results of recent investigations. It is essential that the physician in the nursing home have this information on the arrival of the patient. A routine phone call by the community physician on admission of his patient would certainly enable the physician in the nursing home to deal with any crisis arising shortly after admission. It is very frustrating to have to deal with an elderly patient in an unstable situation and have only several illegible written notes on a government form.

Dealing With the Family

The primary care physician can play an important role in dealing with the elderly and the family during the preadmission period (18). For the elderly person, there is fear and anxiety in the face of the pending loss of independence. Personality and the capacity to adapt to new situations play an important role in how the person will react and what form the reactions will take. The role of the patient in the move, the amount of preparation, and the extent of the environmental change all have an effect on how the person will react to the pending change. Unfortunately, because of the availability of resources, the elderly person generally has little to say as to which institution he/she be admitted to. In spite of these difficulties, preparation and careful follow-up during the pre-admission period can have a significant positive effect (19). This is especially true for the demented patient for whom a move from a stable home environment can be a source of great anguish.

For families, the institutionalization of a parent generally constitutes a crisis. There is a sense of loss, guilt, anger and disagreement among family members which may lead to significant crisis.

The physician must be attentive to the anxiety, fears and guilt which the patient and the families may not express openly. The expression of these feelings, whether in sessions together or separately with the family and the elderly person, should be encouraged. The involvement of a social worker during this period is crucial. Some long-term-care institutions have pre-admission programs to help deal with these problems.

CONCLUSION

The decision to institutionalize an elderly person is an important one from the point of view of the elderly person, the families and society in general. Knowing the patient, his/her environment and the community resources are crucial. Before making a final decision on institutionalization, a complete medical and functional evaluation is necessary. The physician must search for and treat reversible diseases and try to optimize the patient's functional status. If the patient cannot live at home, then an alternative form of housing in the community should be looked at before making a final decision on institutionalization. Once that decision is made, the physician must be active in the pre-admission period and prepare the elderly and the family for admission.

Reproduced, with permission, from Howard Bergman, M.D., Understanding Placement of the Demented Elderly. *MEDICINE, North America,* November, 1989, 4th Series, 1 of 37:94-98.

REFERENCES

1. Schreiber, M. S., Hughes, S. (1982). The Chicago five hospital homebound elderly program: A long-term care model. *Pride Inst. J. Home Health Care, 1*:12-20.
2. Federal Council on Aging (1981). The need for long term care: A chartbook of the Federal Council on the Aging, Washington: U.S. Government Printing Office, OHDS 81-20704.
3. Shamoian, C. A. (1983). *Psychogeriatrics.* Medical Clinics of North America, *67*:361-378.
4. Mortimer, J. A., Schuman, L. M., French, L. R. (1981). Epidemiology of dementing illness. In Mortimer, J. A., Schuman, L. M. (eds). *The Epidemiology of Dementia.* New York: Oxford University Press, 3-23.
5. Larson, E. B. (1989). Alzheimer's disease in the community. *JAMA, 262*:2591-92.
6. Kane, R. L., Ouslander, J. G., Abrass, I. B. (1984). The elderly patient: demography and epidemiology. In *Essentials of Clinical Geriatrics.* New York: McGraw Hill, 17-33.
7. Rubenstein, L. Z., Ouslander, J. G. Wieland, D. (1988). Dynamics and clinical implications of nursing home-hospital interface. *Clinics in Geriatric Medicine, 4*:471-491.
8. Bergman, H., Clarfield, A. M. (April, 1988). Appropriateness of transfer of patients from a nursing home to the acute care hospital; emergency room visits and admissions. Presented to McGill Centre for Studies in Aging, Research Day, Montreal.
9. Kraus, A. S., Spasoff, R. A. Beattie, E. J. (1976). Elderly applicants to long-term care institution. II. The application process, placement and care need. *J. Am. Geriatr. Soc., 24*:165-72.
10. Kleh, J. (1977). When to institutionalize the elderly. *Hosp. Pract., 12*:21-34.
11. Kraus, A. S., Spasoff, R. A., Beattie, E. J. (1977). Elderly applicants to long-term-care institutions: Their characteristics, health problems and state of mind. *I. J. Am. Geriatr. Soc., 24*:117-25.
12. Bergman, H., Clarfield, A. M. (1985). Nursing home admission: When, why, where? *Canadian Family Physician, 31*:2287-90.
13. Kollek, D. (1989). Geriatric stepped care: An organized approach. *Canadian Family Physician, 35*:613-6.
14. Levine, N. B., Dastoor, D. P., Gendron, C. E. (1983). Coping with dementia: A pilot study. *J. Am. Geriatr. Soc., 31*:12-18.
15. Rubenstein, L. (1983). The clinical effectiveness of multidimensional geriatric assessment. *J. Am. Geriatr. Soc., 31*:758-62.

16. Miller, M. B., Elliott, D. F. (1976). Error and omission in diagnostic records on admission of patients to a nursing home. *J. Am. Geriatr. Soc.*, *24*:108-1.

17. Rodstein, M., Savitsky, E., Starkman, R. (1976). Initial adjustment to a long-term-care institution: Medical and behavioural aspects. *J. Am. Geriatr. Soc.*, *24*:65-71.

18. Grant, P. R. (1985). Who experiences the move into a nursing home as stressful? Examination of the relocation stress hypothesis using archival, time-series data. *Can. J. Aging*, *4*:87-99.

19. Shamian, J., Clarfield, A. M., MacLean, J. (1984). A randomized trial of intra-hospital relocation of geriatric patients in a tertiary care teaching hospital. *J. Am. Geriatr. Soc.*, *32*:794-800.

PSYCHOLOGICAL DIFFICULTIES OF CARING FOR DEMENTIA PATIENTS: THE ROLE OF SUPPORT GROUPS

Taher Zandi, Ph.D.

Associate Professor of Psychology
State University of New York
Plattsburgh, NY

INTRODUCTION

Alzheimer's Disease, as Lewis Thomas said, "is the worst form of all diseases not just for what it does to the victim, but for its devastating effect on family and friends."

Policymakers and health care professionals have become increasingly aware that the stress experienced in caring for demented elderly is crippling and could cause psychological and physiological disorders (Montenko, 1989).

Support groups have been identified as an effective form of assistance to caregivers (George et al., 1986; Horowitz & Dobrof, 1982). Similar in nature to social networks, these groups provide a range of assistance, including emotional support, task-oriented information, advice, and a sense of alliance (Brim, 1974; Caplan, 1974).

The dynamics of such groups provide the caregivers with instillation of hope, information, altruism, modeling, sharing the agony of caring for dementia patients and catharsis (Yalom, 1975). The group structure and organization motivate the caregivers to attend and provide opportunities for interaction with the groups (Zarit et al., 1985). The individual members in return contribute to the dynamics of support group and its organization. Cohesion within the group, relevancy of the information disseminated, homogeneity within the group, and physical proximity of the meeting location make the group useful, and thus worthy of attending by the individual caregiver (Mitchell & Trickett, 1980).

A significant body of research concerning the role of social support and the elderly, in particular frail elderly, suggests that those who have access to support systems have a greater survivability rate than those who do not (Coward & Lee, 1985; Scott, Roberto, 1985). Close connections with family members, friends, and other members of the community enhance health and improve psychological adjustment (Dunst & Trivett, 1986; Folkman & Lazarus, 1985). Personal networks provide elderly individuals with companionship, practical help, and useful advice. People with spouses, friends, and extended family are in better psychological health when compared with those with few or no supportive social contacts (Leavy, 1983); this group even utilizes the formal supports more frequently (Stoller, 1988).

New Directions in Understanding Dementia and Alzheimer's Disease,
Edited by T. Zandi and R. J. Ham, Plenum Press, New York, 1990

113

The high level of stress, coupled with inadequate coping resources, frequently lead to various psychiatric symptoms (Klein et al., 1967; Pruncho & Resch, 1989). In addition to the psychiatric difficulties, the caregiver experiences great physical stress (e.g., loss of sleep and incomplete and inadequate nutrition) (Savitsk & Sharkey, 1973). Depression is identified as the main characteristic of the caregiver's psychological difficulties associated with caregiving. The caregiver's level of depression ranges from mild to severe. As Gallagher, Rose, Rivers, Lovett and Thompson (1989) have reported, there are only a handful of studies that have directly examined the prevalence of depression among the caregivers of dementia patients. For example, Coppel, Barton, Becker, and Fiore (1985) investigated the relationships between cognitions associated with coping strategies to depression in spouses (caregivers) of dementia patients. They found a great number of clinically depressed spouses in their sample. The findings of Coppel et al. (1985) dispute the commonly accepted view that age more than caregiving attributes to the elderly caregiver's depression. Furthermore, it has been noted that this unusual number of depressed caregivers is in part associated with the cognitive impairment of the care receiver.

People with a solid network of support are in better psychological health as compared to those with little or no supportive social contacts (Leavy, 1983; Cohen & Willis, 1985). The support network may function as a "cushion" to lessen the impact of emotional changes on the social and psychological well-being of the caregiver. The individual support system may intervene and serve as a buffer between stressful events and reactions of stress by attenuating or preventing a stress response (Cohen & Willis, 1985). This may occur by providing solutions to the stressful event or by reducing the perceived importance of the problem (Lazarus & Folkman, 1984). Availability of social support networks may lessen the perceived harmfulness of the stressor. The caregiver who has access to a support network may sharpen his/her management skills and learn new techniques for dealing with the dementia patient. Individuals who become caregivers of cognitively impaired elderly, in particular, those with irreversible dementia (e.g., Alzheimer's or multi-infarct dementia) are confronted with additional stressors that are not encountered when caring for non-demented elderly. Among them are:

1. gradual decline in cognitive and communication abilities of the patient that cannot be reversed (Reisberg, Ferris, deLeon & Crook, 1982);

2. gradual loss of functional abilities and physical difficulties such as bladder and bowel incontinence (Reisberg, Ferris & Franssen, 1985);

3. gradual loss of self; the patient's mental abilities diminishes to the point of not recognizing anyone including himself or herself (Reisberg, Ferris, deLeon & Crook, 1982);

4. the length of this progressively debilitating disease ranges from 4 to 20 years with each year presenting new and usually greater stressors than the previous year (Chenoweth & Spencer, 1986); and

5. the gradual loss of caregiver's social interaction that declines along with the patient's condition causing further isolation for the caregivers (Zarit & Zarit, 1982);

The typical caregiver of a dementia patient is a female (usually a spouse or a daughter) between the age of 41 and 82 with a high school education (Zarit et al., 1986). The support system of the caregiver, like the informal network of most older people, is dominated by kin, with women assuming the central caregiving role (Tobin & Kulys; 1981, Sangl, 1983). The caregiver, however, is further pressured by other variables besides caring for a frail elderly individual (Brody, 1985). As the aging population has increased, the need for assistance to the elderly, demographic and socioeconomic trends have reduced the availability of family support and caregivers.

Coupled with the increase in the elderly population, has been a trend toward families having fewer children. As the elderly population grows older, so do their children.

Women's (caregivers) increasing participation in the labor force, along with their aging condition, creates greater pressure for those who may also be in need of assistance (Thoits, 1983). According to Zarit et al., (1987), a majority of these caregivers experience burnout and total fatigue, decline in physical and emotional health, and develop feelings of loss of control. The burden of caregiving often generates guilt. The strain and conflict associated with providing care may have a corrosive effect on the quality of the relationship with the older parent (Cantor, 1983), and the resulting lack of affection may create feelings of guilt (Jarret, 1985). Attempting to meet the demands of multiple roles, caregivers often give up time for themselves. Wishing that all the work would end or that they could sleep through the night without interruption also creates guilt. The degree of burnout, guilt, and preoccupation with caring may influence the caregiver's overall adaptability and adjustments (Thoits, 1983).

The caregiver's social life is also curtailed (Sainsbury & Grad de Alarcon, 1970). Caregivers may become isolated and feel alone with caregiving responsibilities (Frankas, 1980). Caregivers of dementia patients express frustration in not finding a solution to their everyday problems. It is very difficult for the caregiver to accept the ongoing deterioration of the patient. The medical and social community cannot provide effective treatment planning and that exaggerates their frustration and increases their burden (Montenko, 1989).

George et al. (1986) suggested that a caregiver's social network declines because of unanticipated changes in the status of the dementia patient. Despite the popular belief, Stoller (1988) found little evidence that networks increase in size in response to greater functional impairment. Zandi & Talmage (1989) investigated the relationship between the caregiver's support network and level of burden and its impact on the patient's cognitive and functional capacities. Zandi & Talmage (1989) reported that the onset of the disease is not as strong a predictor of the caregiver's burden as is the interaction between the caregiver's social isolation and the length of the disease. The caregiver who is fairly isolated from regular social bonds and support has a higher level of burden. The caregiver who is left alone is quite vulnerable to the changes in the status of the patient's abilities. In such situations, the weaker coping abilities of the caregiver influence the patient's cognitive and functional abilities which, in turn, influence the caregiver's burden. Zandi & Talmage (1989) further reported that the level of burden of the caregivers attending support group meetings was less than those who were isolated. As Brody (1985) writes, regardless of how much caregiver daughters do for their mothers, they feel guilty that they are not doing enough. Nevertheless, the rate and the degree to which the caregivers experience these complications varies based on their access to a structured support network (Zarit et al., 1987).

As Zandi et al. (1988), noted, the caregiver who is involved in broader social support networks, as well as specialized support groups, may more effectively manipulate the patient's remaining functional and cognitive abilities, plan better, and, in turn, prevent severe burnout. Another study conducted by Zandi, suggested that the older adults' adjustment processes at the time of stress have to do with their level of involvement with the social and cultural focus around them (Zandi & Talmage, 1989). Nevertheless, in the special condition of caring for dementia patients, if the broader social network does not offer them specific technical support (e.g., respite), its value will be undermined (Zandi et al., 1988).

Support groups by design provide assistance geared toward the particular needs of the dementia caregivers. Persons sharing common problems frequently report the need for more information about the nature of their problems and/or method of coping with it. Support groups are a potentially useful mechanism for transfer and exchange of this information (Yalom, 1975).

Support Group's Functional Role

Zarit, Orr & Zarit (1985) discuss the dynamics and constraints of groups by reviewing the three elements of: 1) group curative factors; 2) group structure; and 3) group norms.

The curative factors are defined as qualities of group interaction, which includes: 1) instillation of hope; 2) universality; 3) imparting of information; 4) altruism; 5) the corrective recapitulation of the primary family group; 6) development of socialization techniques; 7) imitative behavior; 8) interpersonal learning; 9) group cohesiveness; 10) catharsis; and 11) existential factors (Yalom, 1975).

Group structure refers to the organizational aspects of the group, for example, how often it meets, who attends, etc. Group structure, therefore, has implications for group interaction. Finally, group norms refer to implicit and explicit rules that arise out of interaction among participants.

According to Zarit et al. (1985), these variables interact based on the dynamics of the individual members of a group. However, the group as a whole influences individual commitment and determines its cohesiveness. Cohesive networks have a greater chance of surviving and better recognizing its members' needs. The individuals in these groups, because of their social and cultural proximities, can better identify with each others' problems and more effectively disseminate the needed information.

Among the curative factors (Yalom, 1975), those that particularly pertain to support groups of Alzheimer's Disease families are: imparting information, universality, imitative behavior, and interpersonal learning. The variables associated with the group structure factor: group leadership, attendance, and access to the means of attending support groups such as availability of respite while going to a meeting, are very important to the support group structure (Mitchell & Trickett, 1980). Caregivers are often the best source of information. Opportunities should be created for them to express their needs, experiences, and emotions associated with their caregiving problems. This establishes the group norms (Kelly, 1977).
In a recently published article, Toseland and Rossiter (1989) reviewed the literature on group intervention in support of the family caregiver. Their literature search revealed that the majority of support groups included education and support as the main focus of their functioning strategies. The functions of these groups included: 1) information about the patient's situation; 2) the group members emotionally supporting one another; 3) the emotional impact of caregiving; 4) self-care, problematic interpersonal relationships; 5) the development and use of the support systems outside of the group; and, 6) home care skills.

The caregiver's physical health as well as psychological coping abilities were rated high on the list by the groups. The caregivers were assisted in reducing their feelings of frustration by overtly discussing them. The groups also encouraged members to express their feelings of guilt and inadequacy in their caring techniques (Arenson, Levin & Lipkowitz, 1984).

Assisting the Caregiver

Techniques such as stimulation and maintenance of patients' cognitive abilities have also been used as an intervention strategy to relieve the caregiver's burden and the caregiver's level of frustration. Reality orientation training (Folsom, 1968; Zarit et al., 1985) and memory management (Wilson & Moffat, 1984) are only a few techniques associated with the cognitive stimulating model. Quayhagen and Quayhagen (1989) found three stimulation programs (patient-caregiver conversation, memory-provoking exercises and problem-solving techniques) utilized by patients in the program enabling them to maintain their levels of cognitive and behavioral functioning and improve emotionally.

Support for the family caregiver may also be provided on a temporary basis through respite time, respite being the intervals of relief from the demands of the caregiving role. Availability of respite assists the caregiver psychologically as well as physically. The respite in most instances is provided on an informal basis. For example, family members, friends and neighbors provide relief time for the primary caregiver (Stoller, 1984). The caregiver who has access to respite makes more logical decisions, contacts appropriate formal support systems (e.g., Social Services, Alzheimer's Association, etc.) and therefore is less burned out (Montgomery, Gonyea, Hooyman, 1985).

Provision of this type of respite support, although it is commonly practiced, is not readily available to all family. The economic trend, mobility trend, age of the relatives, as well as physical distances, are a few of the sources of such difficulties (Stoller, 1982).

The second type of respite is provided through the use of home health care individuals. This is a paid respite time where the home health provider goes to the patient's home relieving the caregiver of her/his duties of caregiving. This type of support-respite is only effective when there are home health agencies available and when their staff are concerned and well trained.

The major difficulty in obtaining this type of respite is the cost. Given the state of the economy of the elderly population, the caregiver sometimes has to decide whether to spend the money on respite for herself or buy life necessities. Although no systematic data has been collected on the economic impact of the home-health respite, my experience in Northeastern New York supports that this could become a burden itself for those who can barely afford it. According to Lawton, Brody & Saperstein (1989), the cost of caring for a cognitively impaired elderly person at home on a part-time basis ranges between $32-$59 per day. This is a staging source of stress for the dementia caregiver. Caregivers sometimes complain about "teaching how to care" to the part-time caregiver and that the paid caregiver may not provide the quality of care that they could provide.

The third form of support-respite is through Alzheimer's day centers, which are gaining popularity these days. The support and respite provided for the caregiver in this model is twofold: 1) the caregiver is relieved of her regular caregiving responsibilities; 2) the patient's social-emotional status may be upgraded. That, in turn, makes the caregiver's job easier (Holmes, Zandi, Tooke, 1989). The program cost may not be as much as home health care since there are more patients than caregivers. However, the major complaint on this model has to do with preparation and transportation of the patient from home to the day center.

FUTURE DIRECTION OF RESEARCH

Researchers have extensively described caregivers' psychological difficulties associated with caring for dementia patients. This is evident as in 1989 alone, more than 30 percent of the publications in the Journal of Gerontology and The Gerontologist have one way or another studied caregivers' difficulties and burdens (see The Gerontologist, 1989, Volume 29, No. 2 and 4). As Steven Zarit (1989) pointed out, what we need now more than ever before is: 1) to improve research strategies; 2) to develop explanatory model(s) that could shed light on the enormous amounts of data being produced; and 3) we need to distinguish between caring procedures required for the demented patient versus caring procedures required for frail elderly.

Given the economic and mobility trends in our society, families will be pushed further apart. This translates into less direct service delivery contributions by the family. Changes in family living arrangements coincide with the increase in the number of elderly, particularly those above 80 years of age. According to the projected figures of Canadian statistics, by the year 2020, there will be 6.9 to 9.4 million dementia patients and approximately 5.3 million caregivers in the United States and Canada. This is nearly a 40% increase in the number of patients and caregivers of dementia.

REFERENCES

Arenson, M., Levin, C., & Lipkowitz, R. (1984). A community-based family patient care group program for Alzheimer's disease. *The Gerontologist, 24*:339-345.

Brim, J. (1974). Social network correlates of avowed happiness. *Journal of Nervous and Mental Diseases, 158*:432-439.

Brody, E. (1985). Parent care as a normative stress. *The Gerontologist, 25*:19-29.

Cantor, M. (1983). Strain among caregivers: A Study of experience in the United States. *The Gerontologist, 23* (6):597-603.

Caplan, G. (1974). *Support Systems and Community Mental Health: Lectures on Concept Development*. New York: Behavioral Publications.

Chenoweth, B., & Spencer, B. (1986). Dementia: The experience of family caregivers. *The Gerontologist, 23* (3):267-272.

Cohen, S., & Willis, T. (1985). Stress, social support, and the buffering hypothesis. *Psychological Bulletin, 98*:310-357.

Coppel, D. B., Barton, C., Becker, J., Fiore, J. (1985). Relationships of cognitions associated with coping reactions to depressions in spousal caregivers of Alzheimer's Disease patients. *Cognitive Therapy and Research, 9*:253-266.

Coward, R. T., Lee, G. R. (1985). *The Elderly in Rural Society*. Springer Publishing Company, New York.

Dunst, C. & Trivett, C. (1986). Mediating influences of social support: Personal, family and child outcomes. *American Journal of Mental Deficiency, 90* (4):403-417.

Folkman, S. & Lazarus, R. (1985). If it changes it must be a process: Study of emotion and coping during three stages of college examination. *Journal of Personality and Social Psychology, 95*:107-113.

Folsom, J. C. (1968). Reality orientation for the elderly mental patient. *Journal of Geriatric Psychiatry, 1*:291-307.

Frakas, S. (1980). Impact of chronic illness on the patient's spouse. *Health and Social Work, 5*:39-46.

Gallagher, D., Rose, J., Rivera, P., Lovett, S., Thompson, W. L. (1989). Prevalence of depression in family caregiver. *Gerontologist, 29* (4):449-456.

George, L. & Gwyther, L. (1986). Caregiver well-being: A multi-dimensional examination of family caregivers of demented adults. *The Gerontologist, 26*:253-259.

Holmes, C., Zandi, T., & Tooke, W. (1989). Unpublished manuscript.

Horowitz, A., & Dobrof, R. (1982). The role of families in providing long-term care to the frail and chronically ill elderly living in the community. Final report. Brookdale Center on Aging and Hunter College, New York.

Jarret, W. (1985). Caregiving with kinship systems: Is affection really necessary? *The Gerontologist, 25* (1):5-10.

Kelly, J. (1977). The ecology of social support systems: Footnotes to a theory. Paper presented at the symposium, *Toward an understanding of natural support systems*, at the 85th annual meeting of the American Psychological Association, San Francisco, California.

Klein, R., Dean, A., & Bogdenoff, M. (1967). The impact of illness upon the spouse. *Journal of Chronic Disability, 20*:241-248.

Lawton, P.M., Brody, E.M., & Saperstein, A.R. (1989). A controlled study of respite service for caregivers of Alzheimer's patients. *The Gerontologist, 29* (1):9-15.

Lawton, P. M., Kleba, H. M., Moss, M., Rovine, M. & Glicksman, A. (1989). Measuring caregiving appraisal. *Journal of Gerontology, 44*(3):71.

Lazarus, R.S. & Folkman, S. (1984). *Stress, Appraisal and Coping.* New York: Springer.

Leavy, R. (1983). Social support and psychological disorder: A review. *Journal of Community Psychology, 11*:3-21.

Mitchell, R. & Trickett, E. (1980). Task force report: Social networks as mediators of social support. *Community Mental Health Journal, 16* (1):27-46.

Montenko, A. (1989). The frustration, gratification, and well being of dementia caregivers. *The Gerontologist, 4*:166-172.

Montgomery, R. J. V., Gonyea, J.C., & Hooyman, N. R., (1985). Caregiving and the experience of subjective and objective burden. *Family Relations, 34*:19-26.

Pruchno, R., & Resch, N. (1989). Husbands and wives as caregivers: Antecedents of depression and burden. *The Gerontological Society of America, 29* (2):159-165.

Quayhagen, P. M. & Quayhagen, M. (1989). Differential effects of family-based strategies on Alzheimer's disease. *The Gerontologist, 29*(2):150-155.

Reisberg, B., Ferris, S., deLeon, M., & Crook, T. (1982). The global deterioration scale for assessment of primary degenerative dementia. *American Journal of Psychiatry, 139*:1136-1139.

Reisberg, B., Ferris, S.H., & Franssen, E. (1985). An ordinal functional assessment tool for Alzheimer's type dementia. *Hospital and Community Psychiatry, 36*:593-595.

Sainsbury, P. & Grad de Alarcon, J. (1970). The psychiatrist and geriatric patient: The effects of community care on the family of the geriatric patient. *Journal of Geriatric Psychiatry, 4*:23-41.

Sangl, J. (1983). The family support system of the elderly. In R. Vogel and H. Palmer (eds.), *Long-Term Care: Perspectives from Research and Demonstration.* Washington, D.C.: Health Care Financing and Administration.

Savitsk, E., & Sharkey, H. (1973). The geratric patient and his family: Study of family interaction in the aged. *Journal of Geriatric Psychiatry, 5*:3-19.

Scott, P.J. & Roberto, A.R. (1985). Use of informal and formal support networks by rural elderly poor. *The Gerontologist, 25* (6):624-630.

Stoller, E. (1982). Coping with illness: Sources of support for the noninstitutionalized elderly. *Health and Social Work, 7*:111-112.

Stoller, E. (1984). Self-assessments of health by the elderly: The role of informal support, *Journal of Health and Social Behavior, 25* (3):260-270.

Stoller, E. (1988). Informal support networks of the rural elderly: A panel study. Technical report to the National Institute on Aging, Grant Number RO1AG6409.

Thoits, P. (1983). Multiple identities and psychological well-being. *American Sociological Review, 48*:174-187.

Tobin, S., & Kulys, R. (1981). The family in the institutionalization of the elderly. *Journal of Social Issues, 37* (3):145-157.

Toseland, W. R. & Rossiter, C.M. (1989). Group intervention support family caregivers: A review and analysis. *Gerontologist, 29*(4):437-448.

Wilson, B. A. & Moffat, N. (1984). *Clinical Management of Memory Problems,* Rockville, MD: Aspen.

Yalom, I. (1975). *Theory and practice of group psychotherapy,* 2nd edition. New York: Basic Books.

Zandi, T., Talmage, L. (1989). Institutionalized and noninstitutionalized elderly adults' psychological adjustment: An ecological study. *Journal of Psychology and Behavioral Sciences,* 4:110.

Zandi, T., Talmage, L., Zele, D., Arillo, L., & Gaeddert, B. (1988). The role of social support groups in late adulthood. Paper presented at the Eastern Psychological Association Convention.

Zarit, S. (1989). Do we need another "stress and caregiving" study? *The Gerontologist,* 4:2.

Zarit, S., Anthony, C. & Bartselis, M. (1987). Interventions with caregivers of dementia patients: Comparison of two approaches. *Psychology and Aging,* 2:225-232.

Zarit, S., Orr, N., & Zarit, J. (1985). The hidden victims of Alzheimer's disease: Families under stress. New York: New York University Press.

Zarit, S., Todd, P., & Zarit, J. (1986). Subjective burden of husbands and wives as caregivers: A longitudinal study. *The Gerontologist,* 26:260-266.

Zarit, J., & Zarit, S. (1982). Measurement of burden and social support. Paper presented at the meeting of the Gerontological Society of America, San Diego, California.

Zarit, S. H., Orr, N. K., Zarit, N. M. (1985). *The Hidden Victims of Alzheimer's Disease.* New York: New York University Press.

EASING THE BURDEN OF CAREGIVING FOR THE PARAPROFESSIONALS

Lory E. Bright-Long, M.D.

Assistant Professor of Psychiatry
State University of New York at Stony Brook
Stony Brook, New York

INTRODUCTION

Over the last decade there has been considerable literature generated concerning caregiver burden and the effects of intervention (Safford, 1980; Rabins 1982; Barnes, 1981; Davis, 1983; Beam, 1984; Zarit, 1985). This literature focuses primarily on the family as the caregiver and not the long-term care nurse or aide or the home health aide as the primary caregiver.

There are over twenty-two thousand nursing homes caring for 1.1 million Americans (Gwyther, 1988). The high prevalence of mental impairment among elderly nursing home residents makes research and program development in this area a national priority. It is estimated that 60-70% of nursing home residents have a dementia (Brody, 1984), thus we have all been realizing for some time now that we need to direct attention to the value of the role of the para-professional as caregiver. As long term care struggles with the issues of care delivery to an increasingly diverse aging population, the primary, "hands-on" paraprofessional must learn to understand the various processes involved in aging as well as how to interpret whether the behavior of the impaired resident is part of the disease process itself or related to acute, perhaps treatable, illness. Additionally, the paraprofessional caregiver will need to learn to develop relationships with the impaired older person in order to personalize the ever-increasing technical medical care.

With the constant financial constraints and limited staffing pool, the Alzheimer's Disease Assistance Center at the State University of New York, Stony Brook was asked to devise a staff intervention to assist in preparing the nursing aides to understand the who, what, where, why and how of dementia.

METHODS AND PROCEDURES

The intervention took place at St. Johnland Nursing Home, a 200-bed health-related and skilled nursing facility in Kings Park, Long Island. For the past five years a psychiatry consultation service has been established with Dr. Bright-Long through the Department of Psychiatry, SUNY, Stony Brook; a companionship program for the residents is being provided by Community Care Companions; a group "respite" program was developed to relieve staff-caregiver burden and to develop the socialization potential still present in some of those with behavioral problems for depressive symptomatology; an in-service education program for

New Directions in Understanding Dementia and Alzheimer's Disease,
Edited by T. Zandi and R. J. Ham, Plenum Press, New York, 1990

121

nursing departments has been developed and implemented by nursing administration and Dr. Bright-Long; an administrative council was established for monthly discussions of mental health issues; and all of us look for the financial resources to develop both the services and research needed. An outcome also has been that the facility has become a major supporter of LI-ADAC.

In-service education from LI-ADAC began with monthly lectures to the nursing staff on general geropsychiatric issues. The nursing aides requested lectures of their own. Didactic lectures to the day staff nursing aides were tried during the day shift, however, the group was large and found to be at many different educational levels. It was decided to institute weekly one-half hour "rounds" on each of the three units (two skilled nursing and one a mixture of skilled and health related). These sessions were rather loosely organized but focused upon a particular resident or care issue. Each session has 7 aides in attendance. At the same time the evening and night shift requested "live" lectures. These shifts usually viewed videotapes of in-service presentations. It was decided to devise a four-month series of half hour didactic lectures to present to the evening aides (17 in total) and nights (9 in total) and compare general knowledge gained and perception of burden to the more support-like atmosphere of the day shift.

The evening and night shift nursing aides received four didactic sessions entitled "Who is aging?," "What is dementia?," "Why does she act that way?," and "How do I, how do you treat this disease?" These lectures were designed to: 1) present information about normal aging and about diseases that cause dementia; 2) instruct in problem-solving and management techniques; 3) provide an opportunity to discuss feelings, problems and coping mechanisms; and 4) discuss a team-management approach stressing assessment and documenting skills.

Questionnaires were administered prior to the beginning of the support-rounds and the didactic groups. The questionnaires had general knowledge questions, a perception-of-burden section adapted from the Zarit Burden Interview, and a section concerning personal perspective of aging. After four months, the questionnaires were repeated.

RESULTS

All of the nurses, in all the shifts showed an improvement in the general knowledge section. There was an average score of 70 percent correct on this particular section prior to the educational intervention and 95 percent score following the didactic and the discussion/support group format.

Of the ten questions used to assess perception of burden, there was a significant improvement in the overall perception of burden (Figure I). Prior to the series, 25 percent stated that they felt either "a lot" or "too much" burden which correlates to a 3 or 4 out of possible 0-4 scale, whereas after the sessions 66 percent scored in the "none" or "slight" areas (0 or 1 on the 0-4 point scale). The day shift which received the monthly support group format showed a slightly improved perception of caregiver burden. The questionnaire, as structured, did not elicit the daytime versus evening/night variables to support felt from other departments and the difference between the residents on the shifts. The evening and night shift staff, however, were the ones who reported that they felt more support from fellow staff members. Twenty-nine percent of the evening and night shifts reported "always" feeling supported by fellow staffers while 26 percent reported "frequent" support. (Figure 2). The day shift reported slightly less perception of support.

In general, the series seemed to have little impact upon whether the staff felt that the residents were asking for more help than they really needed. Figure 3 shows that before and after the series the staff consistently felt that the residents

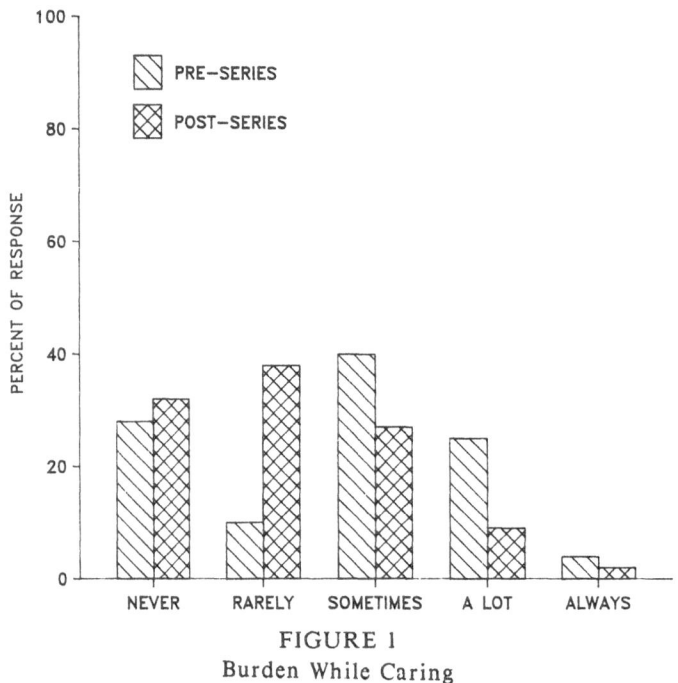

FIGURE 1
Burden While Caring

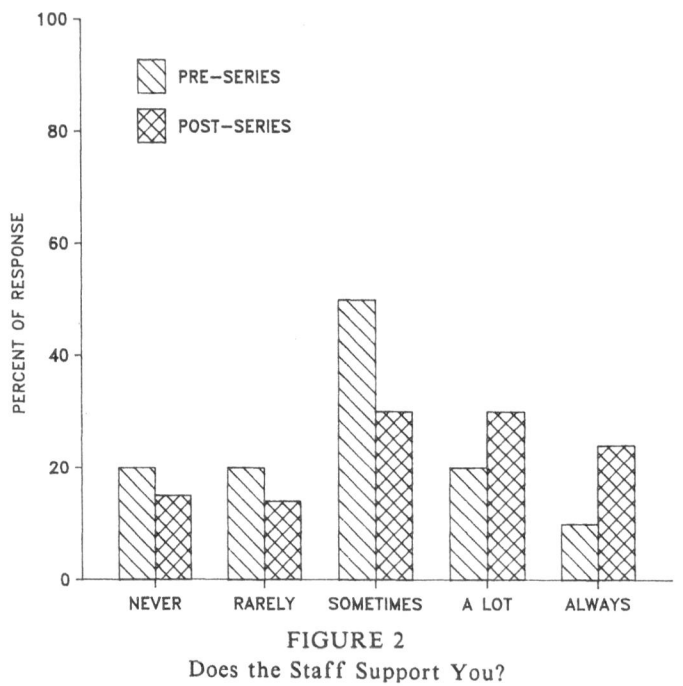

FIGURE 2
Does the Staff Support You?

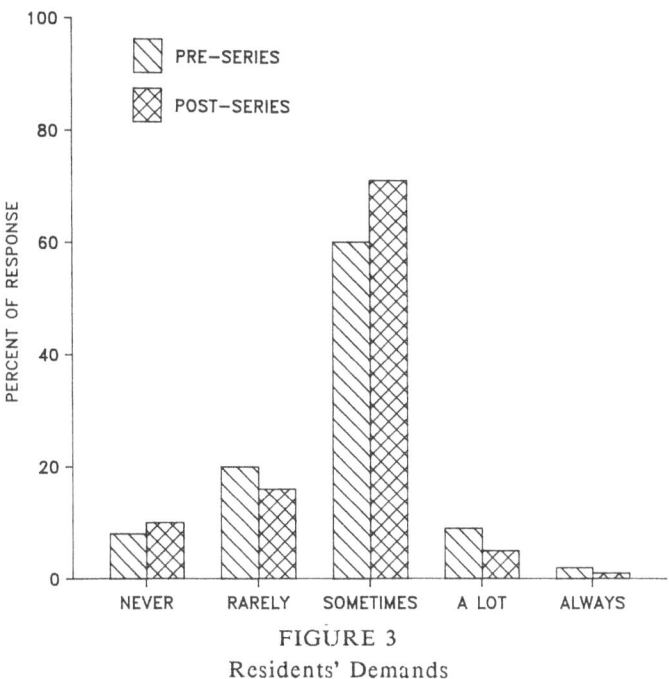

FIGURE 3

Residents' Demands

"sometimes" asked for help when it was not needed. This is a realistic view. However, when asked if the residents expected too much of the staff, 94 percent responded either "never or rarely" after the series. This response changed significantly during the series (Graph IV) perhaps due to the staff's understanding of residents' expectations.

DISCUSSION

The actual intervention of educating the paraprofessional was the goal of this project. The issue of burden and burn-out are real issues. Hare and colleagues (1988) investigated the burn-out phenomenon exploring interpersonal, intrapersonal, and situational variables. The educational materials explored the practical; aspects of intrapersonal and situational variables where the group support sessions focused on the interpersonal and intrapersonal variables.

The development of a therapeutic team was also a goal of the intervention. How can that be measured? The increased interaction of nurses and nursing aides, the increased assessment and communication of findings, the eagerness of aides to participate in facility activities with residents are certainly important indicators. With the sense of decreased burden there is less time spent away from patients in the staff lounge, breaks are frequently taken with residents for a cup of coffee or a cigarette. The request for "PRN" medication has decreased as behavioral interventions are tried and succeed.

Most of current institutional caregiving practices and attitudes, however, will need to be changed if we are to apply even a fraction of what we already know about therapeutically intervening in such maladaptive behaviors as agitation, resisting care, crying out and such disorders as depression, grieving, anxiety, and so

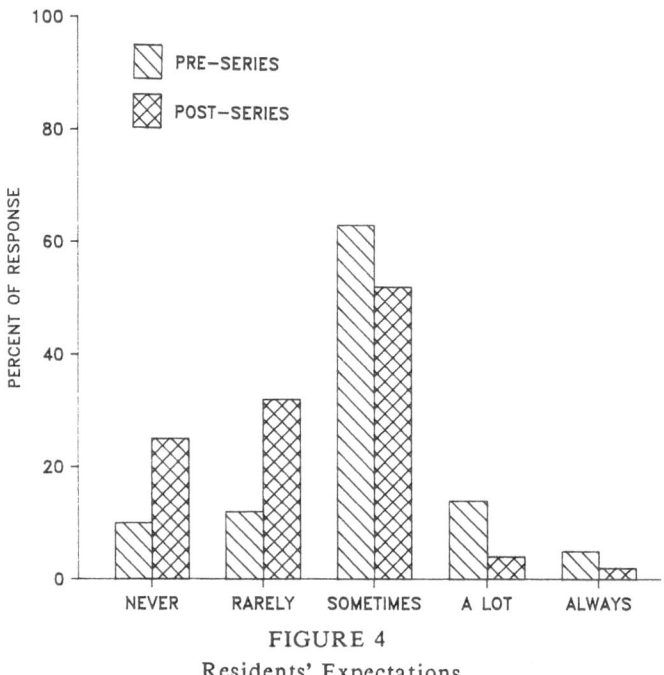

FIGURE 4

Residents' Expectations

on. Preparing and motivating nursing home staff to understand and respond therapeutically to mentally impaired and/or behaviorally disturbed elderly residents will take imaginative, intensive, and flexible education. It will also critically need genuinely committed, courageous nursing home leadership with the will to obtain and direct the necessary resources from the professional mental health community, from academic institutions, and from the public-at-large.

While we are seeing a greater awareness of the need to offer more preparation and a radically improved level of support to paraprofessionals in nursing homes, as late as 1988 over 90% of these "hands-on" people were not provided with any form of continuing education in the area of mental health care. (Cartock, et al., 1988). The reason for this, as we all know, that up until now, the care provided in the "skilled" nursing home was almost totally custodial. The traditional therapeutic pessimism of most nursing homes mitigates against the idea of interactions with the mentally impaired resident that are consciously therapeutic, that is, based on experienced skill and understanding. "Controlling the resident in a kind way," is generally the expected behavior of nurses and aides. This is implicitly assumed to be more a matter of personal virtue and good will than anything that required educational preparation and meaningful support.

Nursing home paraprofessionals are expected to be kind, cooperative, and hard working. Yet most are caught in a conflict between their regard for the resident and the efficiency demands of the institution. Although recent years have seen more studies on the effect of social and psychological burdens on the ability of family caregivers to respond appropriately to the mental health needs of their elderly family member, we have not paid so much attention to the effect similar burdens have on paraprofessionals in nursing homes. We have fairly strong indications that in some areas of the country at least, certainly the large urban cities, most of these nursing home caregivers go to work carrying extremely heavy social, economic, and psychological burdens. As one author puts it, they are either "determined strivers" or "disaffected endurers." (Tellis-Nayak, 1989).

Nursing home paraprofessionals are no longer expendable. There is a critical, national shortage expected to worsen as the consequences of the demographic realities become ever more evident. We will need to develop educational programs of a totally different order from what we were used to; we will need to combine both content, skill, and desire to get involved with the resident; we will need to develop small, focused educational "nuggets" meeting the practical needs as perceived by the participants. The last two years have seen the development of a major state-wide programmatic effort to implement a training program which would provide nursing home paraprofessionals with an understanding of the mental health needs of elderly residents, how to respond to those needs therapeutically, how to enhance and receive group support, and, most importantly, feel valued in what they do.

REFERENCES

Barnes, R. F., Murray, A.R., Scott, M., Murphy (1981). Problems of families caring for Alzheimer's patients: Use of a support group. *JAGS*, *29*:80-83.

Beam, M. I. (Feb., 1981). Helping families survive. *Am. J. Nurs.*, 229-232.

Brody, E., Lawton, M. P., Lebowitz, B. (1984). Senile dementia: Public policy and adequate institutional care. *Am. J. Pub. Health*, *74*:1381-1383.

Cartock, P., Nevins, A., Rzetelny, H., Gilberto, P. (1988). A mental health training program in nursing homes. *The Gerontologist*, *28*(4):503-507.

Davis, J. L. (1983). Support Groups: A clinical intervention for families of the mentally impaired elderly. *J. Ger. Social Work*, *5*:27-33.

Gwyther, L. P. (1988). Nursing-Home Care Issues, In: Aronson, M. K. (Ed.) Understanding Alzheimer's Disease. *ADRDA*, Macmillan, New York.

Hare, J., Pratt, C. C., Andrews, D. (1988). Predictors of burnout in professional and paraprofessional nurses working in hospitals and nursing homes. *Int. J. Nurs. Studies*, *25*:105-115.

Rabins, P. V., Mace, W. L., Lucas, M. J. (1982). The impact of dementia on the family. *JAMA*, *248*:333-335.

Tellis-Nayak, M. (1989). Quality of care and the burden to two cultures: When the world of the nurse's aide enters the world of the nursing home. *The Gerontologist*, *29*(3):302-313.

Zarit, S. H., Orr, W. K., Zarit, J. M. (1985). *The Hidden Victims of Alzheimer's Disease*. New York University Press, New York.

HOW TO CARE FOR DEMENTIA PATIENTS: CASE MANAGEMENT MODELS IN LONG-TERM HOME HEALTH CARE

Elizabeth Pohlmann, R.N., MPH

Director, Alzheimer's Disease Assistance Center
of the Capital Region
The Eddy Center for Aging
Troy, New York

David Howells, S.C.,

Executive Director, Senior Care Connection
Troy, New York

Diane Buchanan, R.N., B.S.

Visiting Nurses Association
Troy, New York

INTRODUCTION

Caring for dementia patients is a task that requires planning, preparation and special dedicated people. Today, in our society, we have just started learning how to develop these specialized care delivery systems. This is an awesome task and will require cooperation of all people involved in dealing with the dementia patients.

The time is ripe for both state and private organizations to recognize the need of the dementia patients and their family caregiver. Today, more than 80% of the caring for dementia patients is being accomplished through the efforts of the family members. The remaining number of dementia patients either live alone or live in nursing homes.

There is still a great amount of confusion among state and federal agencies as to what office should oversee the affairs of the dementia patient. For example, recently the State of New York Office of Mental Health refused to recognize dementia as a mental health problem. The nursing homes are not interested in these people because they are difficult to care for and financial reimbursement through Medicaid for these patients is less than that which Medicaid pays for patients requiring skilled nursing care. This situation is alarming for nearly four million dementia patients in the United States and Canada and another four million caregivers. Ironically, these caregivers are mostly spouses and daughters of the patients who are often, themselves, in need of medical services, both physical and psychological.

In the next two chapters, the authors have provided answers and examples of how the private and government sectors have been responding to the needs of dementia patients and their family members.

New Directions in Understanding Dementia and Alzheimer's Disease,
Edited by T. Zandi and R. J. Ham, Plenum Press, New York, 1990

Elizabeth Pohlmann, David Howell and Diane Buchanan demonstrate in their chapters how a multidisciplinary case management program is formed and how the case management program in partnership with social services and long-term home health care programs are cooperatively lessening the burden of care and sharing the anguish of the caregiver. In Chapter 13, Suzanne Lavin, program director of a mobile geriatric team, an innovative multidisciplinary task force program, will disclose how the multidisciplinary effort can function as case finder of endangered, socially isolated, older individuals with major psychiatric problems and older people threatened by the cognitive impairment of dementia. Lavin uses case studies to describe how the "home visits"--what she refers to as the "primary tool"--is to be used in detecting, planning and managing an elderly dementia patient's life situation.

PRIVATE SECTOR CASE MANAGEMENT--THE SENIOR CARE CONNECTION

Background, Evolution and Development

In July 1988, the Senior Care Connection, a private not-for-profit organization began operation. This came about after years of discussions and deliberations. Through its institutional base and community service programs, The Eddy Family of Services for the Elderly sought to create an entity that could help persons in crisis who came for help: a spouse or parent whose health had deteriorated to the point where placement in a facility or care at home for 4 to 24 hours was necessary. Of course, responding in a crisis with 2 or 3 days notice was usually impossible.

In many situations, the Alzheimer's victim may be suffering from other co-morbid disorders. The care-providing agencies are just now making efforts to provide a broad range of assistance. Following is an example of such a model.

Senior Care Connection is designed to provide care management services to an elderly or disabled individual whose situation falls into one of these categories:

- Frail health, living alone or with frail spouse, no
  family involved locally
- In need of services from multiple entities
- Multiple impairments, without established service
  provider
- Independent, but concerned about future
- Interested in a link to a multi-level, multi-service
- Geriatric care network without immediate service needs

The goals of the model program are to offer services to individuals who were not in a service delivery program and to link services more closely so they could be delivered through a single entry point. Considering the caregiver's level of burden, it is not only efficient to funnel services into one broker rather than have the caregiver run around to find different needed services, it can also be therapeutic.

This venture into care management is not without precedent and research into past and current models. Following are examples of models that assisted the evaluation of multidisciplinary care units.

The National "Channeling" Long-Term-Care Demonstration Program, Rensselaer County, New York was 1 of 10 sites nationally which focused on case management coupled with service delivery and monitoring. Existed from 1982-1984.

Connecticut Community Care, Inc. Offers care management to private individuals along with its core services to clients of publicly supported programs.

The services offered by private geriatric case managers were studied. Services for Independent Living, Inc. (SMILE) is a Robert Wood Johnson-funded initiative operated in Albany, New York and providing case management and home care services to the independent elderly. Services are provided directly and through contact. Various programs that focus on benefit control, commonly referred to as case managers, whose focus is benefit administration/service linkage.

These programs are representative of the primary private care management models:

(1) Private practice (solo or group)
    usually a social worker or nurse
(2) A gatekeeper/benefits control case manager
(3) A provider-linked private non-profit corporation

Case management programs may differ from one another depending on the type of services they provide. For example, the Senior Care Connection, Inc. differs from the provider- linked model in which this program, while linked to a network of services, is not directly involved in home care or nursing home services. In addition, the Senior Care Connection, Inc. is not limited by contract to a select few service providers. It is a broker of services regardless of the source. This we felt was an essential flexibility to maintain. Shortages of manpower and nursing home beds meant that our success in arranging services would be seriously compromised if we limited ourselves to a small group of providers under contract. Thus, we now deal with all providers in the region regardless of sponsor or affiliation.

The majority of the providers in the care management model provide the elements of care management outlined by the National Council on the Aging. These elements are part of the program:

Intake/Screening
Assessment
Care Planning
Monitoring
Discharge/Termination

However, provision of the elements of care management may vary depending on the nature of the care management model.

The first step for starting a care management program is to understand the consumers' perception of care and their needs. For example, the Senior Care Connection of Albany, New York began a two-phase market research during mid-1987 through early 1988. Phase 1 consisted of a series of four focus groups with these groups of people:

The elderly
Caregivers - adults with aging parents
Professionals - lawyers, accountants, physicians
Selected senior management staff of the EDDY program

Here they learned formally that this service is easier to introduce to the caregiver than it is the aging elderly. Other responses indicated a high level of acceptability, a willingness to incur expenses for these services, and provider affiliation. Conflict of interest is not a problem.

During phase 2 they distributed a ten-page questionnaire to 79 elderly, 74 caregivers and 21 professionals. They were asked their reaction to the general concept and to specific services to be provided. Their responses showed strong interest in such a program. Specifically, these were some of the highlights of their responses:

1. Stays in touch
   via weekly/monthly checks and monitoring
2. Helps deal with the bureaucracy
   advocacy + assistance with "red tape"
3. Promotes independence
   responds to needs, provides personal attention
4. Easily and immediately accessible
   24 hour access

Findings showed no substantial differences between the caregivers and the elderly in these responses. On the question of the willingness to purchase services, the key feature in both groups was the arrangement and coordination of services. Responses of the professional group were also similar to these responses.

The program was then started in July, 1988 using the findings mentioned above as the guidelines. Since then, they have provided information and referral service to hundreds of individuals averaging 30 requests per month. The program has assessed and served 78 individuals of whom 54 have joined the program. Of these 54, 10 have been discharged due to:

Death	2
Referral to adult protective	1
Felt service no longer needed	1
Placed in a facility	6

The most significant referral sources used by this program were:

Hospitals
Provider organizations
Advertising
Employee assistance programs
Attorneys

Since the implementation of this program, the evaluation has revealed that:

The program is extremely difficult to market
It is a new concept that is somewhat abstract and attractive primarily to:

Caregivers who have experienced some difficulties
Professionals involved with an elder
Frail elderly individuals interested for a
   spouse or recommended by a provider

Pricing is extremely difficult to set for a "break-even" goal. There is a huge gap currently between the cost of delivering this service and the perceived value. This is something that we expect to improve as the public is better educated on this service and the program matures.

Finally, the cohesiveness of the program is the key element in the vitality of the program.

Another example of a case management program that is also implemented in the Capital Region of the State of New York is the Alzheimer's Regional Management Services (ARMS). This program is funded by the New York State Department of Health and is awarded to the Visiting Nursing Association (VNA) of the Capital Region.

ALZHEIMER'S REGIONAL MANAGEMENT SERVICES

Alzheimer's Regional Management Services known as A.R.M.S. is a program of the Visiting Nurse Association of the Capital Region, Troy, N.Y. under a grant from the NYS Department of Health. It is currently in its third year of operation and covering five counties: Albany, Columbia, Rensselaer, Saratoga, and Schenectady. The purpose of the program is to provide diagnostic assessment and case management services to patients with Alzheimer's disease and other dementias and their families and to enable the family to keep the patient home as long as possible.

The proposal of the A.R.M.S. program began due to the Visiting Nurse Association observing the unmet needs of patients with dementias and having state monies available as a funding source. The requests for service that the V.N.A. was receiving from the dementia patients and their families were not Medicare reimbursable. In an effort to prevent these patients from continuing to "fall through the cracks," A.R.M.S. was developed. To the present, A.R.M.S. continues to receive NYS Department of Health funding via a grant, funding from the V.N.A., and in kind services of approximately 10%.

Participation in the first two months of A.R.M.S. includes the assessment, family conference and home visit components of the program. The assessment is provided by a geriatrician, nurse coordinator and social worker. In addition to diagnostic determination via assessments, needs are identified for the patient and family. Families receive assessment findings, diagnosis, needs and case management plan at a subsequent family conference with the team members approximately one month from the assessment date. A home visit is scheduled at the conclusion of the family conference in order to complete an environmental assessment and to review any questions that there may be concerning the information provided at the family conference. Primary physicians also receive a copy of the A.R.M.S. physical findings, lab studies/testing as pertinent, and the findings, diagnosis and plan presented to the family.

After the home visit is completed, patients and families have the opportunity to continue with the A.R.M.S. program through the ongoing case management component. Ongoing case management includes a reassessment home visit once each six-month period to evaluate the patient's functional level and discuss with families any concerns they may have. Recommendations are made regarding the needs that are present at the reassessment of the patient. The primary physician as well as the family receive a copy of this reassessment. Families also have the availability to call and speak with the nurse coordinator or social worker about their concerns Monday through Friday during work hours. The nurse coordinator and/or social worker also make scheduled calls to check on patient status approximately every other month.

There are fees for the A.R.M.S. program. Currently, the initial assessment fee for the first two months of the program is $120, which does not include the physician physical or any testing ordered. The ongoing case management fee is $80 for each six-month period.

The outgrowth of the needs is the basis for the case management plan which is developed. Case management can address areas of:

- self care
- home maintenance, safety and legal issues
- mobility
- elimination
- nutrition maintenance
- sensory/perceptual aspects, visual and/or
    auditory impairments, hallucinations
- short term thought processes
- communication
- socialization

- emotional integrity
- behavioral management, agitation, wandering
- family coping potential, future planning
- other categories as indicated

The appropriate needs for each patient are then viewed in the case management plan with regard to reducing those needs. This may include recommendations for:

1. Informing families of supportive services in the
     community. These can include:

    -Alzheimer's Association support groups
    -O.R.E. volunteer, educational information
        via Alzheimer's Association
    -aide agencies
    -adult daycare
    -legal consultation
    -respite facilities
    -counselors
    -meals-on-wheels programs
    -van transportation via specific hospitals or daycare, etc.
    -talking books for visually impaired

2. Communication strategies which families can employ
     in order to improve patient and family
     understanding of a message. These can include:

    -use of simple language
    -establish eye contact
    -breaking down of tasks
    -use of one step direction
    -allowing the patient time to process the message
    -repeating your communication if there is no response
    -use of prompts and cues as indicated
    -initiating a task as appropriate for patient
    -redirecting patient to a task as indicated

3. Structuring of behavioral management approaches and reducing agitation levels. These approaches can include:

    -calm, supportive atmosphere
    -being aware of your emotional tone
    -good communication strategies
    -establishing regular routines concerning bathing,
        dressing, toileting, and eating
    -avoid confrontation, arguing with patient
    -avoid use of negatives
    -use of non-verbal communication
    -use of distraction
    -use of appropriate sensory stimulation
    -use of medication intervention as deemed appropriate

In viewing the program, there are additional areas of concern which are beginning to be addressed or will need to be addressed in the future whether through A.R.M.S. or another vehicle. These include:

- The difficulty in reaching bedridden patients to whom it may be
   too difficult to come for assessment.
- The difficulty in transporting a patient for an hour or more drive
   from outlying areas in order to be assessed.
- A.R.M.S. does not currently serve a large portion of the individuals

on Medicaid. This may be due to referrals being made to the Medicaid reimbursable services such as medical daycare or LTHHC.

- A.R.M.S. is in the process of merging with the ADAC in a natural progression of being able to offer a specialized diagnostic and case management component as well as educational, information and referral services. This means that the A.R.M.S. program will be available in 10 counties after July 1, 1989.

REFERENCES

American Association of Retired Persons. (September, 1987). More firms aid workers who care for older relatives. *AARP New Bulletin.*

Bureau of Ambulatory Reimbursement, Office of Health Systems Management, NYS Department of Health, (July, 1986). Case management time study: Long-term home health care program.

Connecticut Community Care, Inc. (1986). Creating opportunities for independent living.

Connecticut Community Care, Inc. (1988). Northeast utilities system employee caregiver survey: Findings and observations.

Creedon, M. A. (ed) (1987). Issues for an aging America: Employers and eldercare--a briefing book. The National Council on Aging, Inc.

Creedon, M. A. (Nov.-Dec., 1988). Why are more and more employers taking eldercare needs to heart? *Perspective on Aging.*

H. Linn Cushing, Inc. (April, 1988). Final report: Market analysis of case management for the elderly.

Dawes, S. S. (1986). New York State Project 2000 Report on Long-Term Care. Nelson A. Rockefeller Institute of Government.

Grisham, White, Miller (1983). Case management as a problem solving strategy. *Pride Institute Journal of Long Term Health Care,* Vol. 2, No. 4.

Mature Market Report. (July, 1988). Caregivers mean business.

National Council on Aging-National Institute on Community-Based Long Term Care. (March, 1988). Case Management Standards-Guidelines for Practice.

Paine, T. H. (1987). Employee benefits in 1995. Hewitt Associates.

Parker & Secord. (October, 1987). Private geriatric case management. *DRG Monitor,* Vol. 5, No. 2.

Secord, L. J. (July, 1987). Private case management for older persons and their families--practice, policy, potential, interstudy.

Sommers, T. and Shields, L. (1987). *Women Take Care; The Consequences of Caregiving in Today's Society.* Triad Publishing.

Stein, E. (June, 1988). Close-up: Seniors plus case management for Elders. *Provider.*

Strickland, T. and Boling, J. (April, 1989). "Win-Win" case management assures appropriate medical care. *Continuing Care.*

Zitter, Mark. (March 24, 1989). Don't shy from seniors--they're your future. *Modern Healthcare.*

ENCOUNTERS OF THE MOBILE GERIATRIC TEAM

Suzanne Lavin, RN, MS

Program Director, Mobile Geriatric Team
Mohawk Valley Psychiatric Center
Utica, New York

The mobile teams were organized by the N.Y.S. Department of Mental Hygiene in 1968 (Cumming's memo #68-27) to prevent inappropriate admission of elderly people over age 65 years to state hospitals. Admission guidelines were established around the concept of treatable, reversible conditions, and in-patient admission was to be considered only if documented psychiatric illness was present and could not be managed in another type of community setting. The Mobile Geriatric Teams were based at each state hospital and after twenty years in existence they still remain the touchstone of the present day admission policy for geriatric patients.

In 1978 under the direction of the Unit Chief of an In-patient Geriatric Unit, the MVPC team was reorganized and staffed on a full time basis. The same criteria for admission and the same directive for the team composition and function were used. By this time the name changes were that of the Department of Mental Hygiene, which became the Office of Mental Health, and the state hospitals, which were renamed psychiatric centers.

The functions of the present Mohawk Valley Psychiatric Center team have expanded far beyond that of simply screening for admission. In large part the MVPC team's original function has grown into consultative and educational areas, as well as organizing and coordinating a local chapter of the Alzheimer's Association and its three area support groups.

The team is always mobile and the person is seen in his own familiar setting. Home evaluation is known to be more comfortable, easier on the older person and the caregiver, and far less threatening (Selan & Gold, 1980). Often it is the only means of evaluating and initiating treatment for the person who refuses to leave the home to seek treatment which could minimize or diffuse disruptive symptoms and prevent premature or inappropriate institutionalization.

The team's outreach of "in-home" evaluation and consultation with other agencies and medical services is available to individuals who reside in the geographic area served by the State Psychiatric Center.

The team consists of community mental health nurses, psychiatric social workers, a full-time clinical physician, a part-time OMH psychiatrist and a psychiatrist from the community who acts as a consultant and meets regularly with the team. The team conducts a one-or-two visit comprehensive psychosocial assessment

New Directions in Understanding Dementia and Alzheimer's Disease,
Edited by T. Zandi and R. J. Ham, Plenum Press, New York, 1990

135

and formulates a treatment plan which links the older person to treatment services in the community.

Besides the home visit, the team approach is another valuable tool of the MGT. Visits to first time or new cases are never made by a single team member. Usually two and, if need indicates, three team members go on a home visit. This has many advantages: 1) there are two people to listen to and evaluate the situation; 2) development of rapport with the older person and the significant family makes the process much easier, and the participation of two people makes this twice as likely; 3) two people provide two distinct personalities to experience empathy, and often this situation enables the older person and the family/caregiver to separate and tell their stories.

Informal follow-up contacts support the concept that most older people do improve or stabilize so that they can be maintained in the community. Those who eventually require in-patient admission do so when all the social supports and community treatment modalities can no longer be kept in place for that individual (Wasson, Ripeckyj, Lazarus, Kupferer, Barry and Force, 1984).

Over the past ten years (1979-1989) referrals to the Mobile Geriatric Team have steadily increased. The most recent annual report of 679 individuals referred resulted in an in-patient hospital admission for approximately (3%), usually for short-term stabilization or in some severe situations, for long-term chronic care. Often if the individual is able to be a voluntary admission, and if a brief hospitalization is indicated, the team has been successful in referring them to one of the area's general hospital psychiatric units. An important part of the MGT's function is to make admission to the psychiatric center, when it is needed, as atraumatic as possible for the older person. Including the team as part of the Geriatric Unit under the direction of a common unit chief, ensures continuity and makes this possible. Pending admissions, criteria for discharge, history and background, in-hospital progress, and discharge plans can be discussed regularly by in-patient and out-patient staff.

Referrals to the team are made by physicians, families, friends, neighbors, long-term-care facilities, and community agencies. Recently there is an increasing trend toward self referrals from older individuals who are sensing that they are having increased problems.

The living situations of those referred are usually divided into three groups: (1) that of people living alone; (2) those who live with family or friend; and (3) those who already reside in long-term care or supervised settings. There is usually an increase in referrals at holiday times and during the vacation months when adult children living at a distance visit and become alarmed at marked changes in an older person. A contrast to this is when a family member notices marked deterioration, but it goes unreported or unevaluated because the changes are accepted as "just the results of old age." Reversible problems are sometimes mistaken or written off as "senility." Often the changes are insidious and the older person is able to cover up much of his/her deterioration by seeking out or staying with the familiar and using ingrained social skills. At some point these individuals come to the attention of the MGT because some unprepared for or unanticipated crisis occurs in their living situation.

The circumstances that I would like to present in this section are cases of individuals with quite severe dementia of the Alzheimer's type (DAT) often found living in very dangerous situations, often undiagnosed and untreated. In these examples of cases evaluated by the MGT you will see that there are multiple reasons why individuals often become quite deteriorated before they are referred.

Case #1

Older people are often invisible and isolated. Many times they have no close contacts to witness the deterioration in functioning and they are often accepted as they are, or are past the point of being able to get help for themselves. Lifetime anti-social behaviors of crankiness, suspiciousness, imperiousness, and general unkemptness keep people from noticing a great deal of difference in this population. They continue to rule the household while dangerously impaired; they lack insight and judgment and yet make decisions affecting themselves and others.

        Mr. and Mrs. F were an elderly couple referred by their daughter, an only child. She was mostly concerned with her mother's anxiety and depression caused by her disappointment <u>at living out her retirement years with a cold domineering husband described by both mother and daughter as "impossible</u>." Mrs. F had retired at age 65 and used her retirement money to buy herself a small red sports car--she drove it happily for 10 years until at age 75 her health began to fail. Her husband made an offer to sell it to a neighbor without consulting her. When the neighbor offered to back out, he said he would only arrange to sell it to someone else if he did not take it. Mr. F was becoming quite forgetful, but was still very mean and sarcastic. His wife and daughter were sure that nothing would ever change. He continued to rant and rave and make all household decisions. He left the house on several occasions in his car and became lost and had periods when he did not recognize people. He also talked of "strangers around the house," one, a woman whom he did not recognize. One night as he roamed about the house he came upon his wife who had gotten up to check on him. He struck her on the head and she fell down several stairs. Mr. F then called the sheriff to report that a strange woman was in his house but that he had "fixed her." Mrs. F died of her head injury two days later. Mr. F was arrested, evaluated in a forensic unit, and determined by some MD's to be severely demented--others felt that he was not so demented and covering up. At 81 years he was determined by the judge to be too impaired to be tried for manslaughter. He remains in the OMH Forensic Unit without trial or sentence.

Case #2

        There often exists in the life of an older person a group of "natural helpers" or helping networks that enable them to function in spite of a severe cognitive impairment. A family member or friend visiting in the home might not even be aware of all the daily helpers and contacts that keep an older person "out of trouble." Here is a partial list described by Hooyman and Lustbader (1986) in their book <u>Taking Care: Supporting Older People and Their Families</u>.

        Nurses and receptionists at medical offices
        Bus drivers
        Apartment managers
        Grocery story clerks
        Postal carriers
        Bartenders
        Meter readers
        Pharmacists
        Cab drivers
        Hairdressers and barbers
        Restaurant staff
        Neighbors
        Home health aides

        Family members should not discourage these helpers but should stay aware of how they are providing support, know their limits and thank and acknowledge their assistance whenever possible. Other people won't tell you about these helpers; the human instinct is to avoid acknowledging dependency.

Mr. M was an 85-year-old man, divorced at age 65 and living in a single room with kitchen privileges. He had been referred to the MGT by a neighbor who was wearing out from monitoring Mr. M through several nights of confusion and wandering. Two other tenants, one handicapped in a wheelchair, kept an eye on him and reminded him to go for meals. A local coffee shop considered him a "regular" and served him appropriate meals that he did not have to order for himself. He frequently became locked out of the building, but the handicapped neighbor was usually there to open the door. He continued to wander and began to go to the local police station to report visual and auditory hallucinations which had a paranoid flavor and were increasing with more agitation. Mr. M looked particularly neat and clean because of a lifetime habit of polishing his shoes and keeping his clothing in good order. This good appearance made him seem to be functioning much higher than he actually was. Several friends from his Legion drove him to meetings but did not know of his wandering and confusion. The postal carrier told MGT staff of mail Mr. M often received from out of state, stating it was from a granddaughter. MGT staff worked with adult protective workers in contacting relatives and initiating nursing home referrals.

Case #3

A more complex situation that causes impaired older people to be missed in getting services is the newer phenomenon of step-families (Hooyman and Lustbader, 1986). Adult step-children and step-siblings often pose distinct problems in getting help for older step-parents. There are few established patterns or norms about how middle age children relate to older step-parents. Often families are unhappily aware that their own parent is in the caregiving role, and there is resentment to "my mother's husband" or "my father's wife" who needs the care. Accusations such as "it's your parent who is sick and dragging mine down," or a telephone call in the middle of the night saying, "Come and get your father," are often heard between step-siblings. Distance, personality clashes, unresolved resentment, old loyalties, as well as extensive financial handouts, interfere with someone stepping in, taking charge, and referring an older person for evaluation and intervention until it becomes a crisis situation. Step-families will continue to carve out their healthier roles and relationships in our society. Some are very positive and the families have blended and grown together, but it is still the negative situations that attract our attention to the older individual, often too late to intervene with treatment.

Mrs. M is a 69-year-old widow residing in public housing reported to have severe memory loss and deterioration associated with Alzheimer's type dementia. Management reported that she never ventured outside of the apartment, but reportedly sat on the couch most of the day watching television. Home delivered meals were delivered daily by volunteers who were becoming concerned by her heavy smoking and burned holes near the furniture and in the rugs. Her bills were paid regularly and she received daily short visits from her step-daughter, who had some affection for her, but had very heavy demands from her own family. After much urging and negotiating, we managed to have the step-daughter and the apartment manager present during our assessment. The step-daughter acknowledged that she used the cigarettes as "a babysitter," feeling that as long as she didn't run out of cigarettes, Mrs. M would stay quietly in the apartment. The three rooms were heavily saturated and stained with cigarette smoke and Mrs. M was now smoking "4 to 5 packs each day." Because of the smoking hazard, MGT arranged for a *PPHA to evaluate her for admission to that supervised setting. We

---

*PPHA= Private Proprietary Home for Adults with special supervised area.

pointed out the need to carefully supervise her smoking until she could be moved. The step-daughter addressed this problem by bringing <u>no more cigarettes</u>. When Mrs. M asked for a cigarette, she simply said, "Don't you remember that you don't smoke anymore," and miraculously Mrs. M accepted this statement as fact.

Case #4

Older individuals who are gay or lesbian and have lived for years in the same sex relationship in shared homes face a dilemma when one of them becomes severely demented. Many times they are estranged from their own families and they find it wrenching to have to face up to the frail condition of a partner. It requires that they address all of the caregiving issues that a heterosexual couple would have to face as well as more complex issues. Questions to be determined involve: 1) Will they be separated? 2) Who gives consent? 3) What income will be spent? and 4) Will the caregiver partner be acknowledged and supported as the caregiver or will they encounter non-acknowledgement by the medical profession? The turmoil of not being the legal spouse often keeps the functioning partner "hanging on for awhile longer" in spite of much deterioration and dangerous acting behavior in the household. As the medical community and the helping agencies break away from the stereotypes and work more comfortably with these couples, they can help steer them toward sound legal advice regarding durable power of attorney, wills, living trusts, and rights of survivorship bank accounts (Hooyman and Lustbader, 1986).

Miss K and Miss E were retired nurses who had lived together and worked together for over fifty years since they met in nursing school. They shared a lovely home together and had a deep personal relationship. They were both quite introverted and lived rather isolated lives with only a few close contacts over the years. They always promised to be there for each other for the rest of their lives. They enjoyed retirement together, traveled and enjoyed reading and gardening together. At age 72 Miss K became progressively more and more demented and required constant supervision. When Miss E became worn out, she obtained a nursing home placement for her friend. On the day of placement, Miss K was more alert than she had been in months and said to Miss E, "How could you do this to me?" The next day Miss E removed her from the nursing home, hired some part time help, and struggled for nine more months with the 24-hour care. MGT staff supported, counseled, and assured Miss E while Miss K was admitted to a Day Program. Once she adjusted to being out of the home, the permanent nursing home placement was made and Miss E visits her daily.

The staff of the Mobile Geriatric Team knows the value of the home visit. Assessing older people and their significant caregivers in the home setting is one of the most effective methods of intervention. Networking with other agencies is another valuable tool. The MGT has been a leader and a catalyst in organizing joint visits with multiple agencies such as visiting nurses, family practice physicians, adult protective caseworkers, city and county outreach workers, screening teams from long-term care facilities and other specialized service providers. A visual examination of the older person and the particular situation can be very effective in allowing agencies to decide what their involvement can be based on their agencies' expertise. Working in a collegial manner with other aging service providers helps to build a network of helpers. It enables an agency to establish its own credibility, avoids duplication of services or turf issues, and enables helpers to identify gaps in services and to concentrate on filling those gaps. It prevents any one agency from being overwhelmed, and it helps to prevent staff and agency burnout because there is much satisfaction in being part of a successful intervention.

REFERENCES

Cohen, D. and Eisdorfer, C. (1986). *The Loss of Self: Family Resource for the Care of Alzheimer's Disease and Related Disorders.* New York: W. W. Norton.

Cumming's Memo No. 68-72. Issued by New York State Division of Mental Hygiene, John H. Cumming, M.D., Deputy Commissioner of Mental Health.

Hooyman, N. R. and Lustbader, W. (1986). *Taking Care: Supporting Older People and Their Families.* New York: The Free Press.

Salen, B. and Gold, C. (1980). The late-life counselors' service: A program for the elderly. *Hospital and Community Psychiatry, 31*:403-406.

Wasson, W., Ripeckyj, A., Lazarus, L. W., Kupferer, S., Barry, S. and Force, F. (1984). Home evaluation of psychiatrically impaired elderly: Process and outcome. *The Gerontologist,* Vol. 20, No. 3, pp. 238-242.

Zarit, S. H., Reever, K. E., Bach-Peterson, J. (1980). Relatives of the impaired elderly: Correlates of feelings of burden. *The Gerontologist, 20*:649.

ROLE OF SUPPORT GROUP FOR THE FAMILY CAREGIVER OF DEMENTIA:

RECENT DEVELOPMENTS IN THE STRUCTURE OF THE SUPPORT SYSTEM

Alice Barbara Vickers, R.N.

Co-Founder, Alzheimer's Disease and Related
Disorders Association
Capital Region
Albany, N.Y.

CAREGIVERS AND SUPPORT GROUPS

Growth in General

In the last few years there has been a growing interest in self-help and a proliferation of such groups for various reasons. Some are associated with everyday living, such as the Parent and Childbirth Education Society, and others have a special interest, such as the Epilepsy Foundation of America. Some groups are led by peers and others by professionals, some are for parents or siblings and others for people with special needs. Several states and many cities and counties have developed clearinghouses for information about such groups. Many organizations, such as the American Diabetes Association, have developed their own self-help groups and materials on how to do it. The Family Survival Project (FSP) has served the San Francisco area since 1977 with information, support groups and advocacy about chronic brain disorders and head traumas which occur in adults. One of the pioneers of the Alzheimer's Association was a caregiver, Bobbie Glaze Custer (1), who personally called people all over the country in order to help them with advice and help in establishing groups, thus establishing a network of caring people. She refers to her own personal story as "like a funeral that never ends."

Definition

Most groups in the Alzheimer's Association are open-ended, social support groups, as opposed to closed. A closed group is more formal; a certain program of information is given over a planned period of time, and the group members are asked to commit themselves to attending all sessions (2). In the open-ended groups, the members attend when they feel the need. It has been noted that social support does not always help dementia caregivers who may be looking for help and respite for themselves, only to find themselves embroiled in problem-solving for others in the group (3). Also, it is important for the leader to recognize when someone needs individual professional counseling, especially if a member is depressed. Some people prefer to talk one-on-one, and a list of professional counselors should be available. It is generally recognized that a support group may not serve everyone's need (4).

New Directions in Understanding Dementia and Alzheimer's Disease,
Edited by T. Zandi and R. J. Ham, Plenum Press, New York, 1990

141

Plight of AD Caregiver

Caring for a demented person at home is both stressful and burdensome. The primary caregiver may become another victim of the disease, especially if the caregiver is an elderly spouse with health related problems, or even a parent for whom the physical effort and long hours may be overwhelming. Women are most often the caregivers. Commonly, children who are caregivers may have the added psychological burden of role reversal. Many articles and books have been written about the problems associated with caring for parents. However, it is important to leave some decision-making and control to people with AD as long as possible in order for them to keep their self esteem. Both home and institutional care are expensive, the rewards of caring are few, and the outlook is gloomy because there is no improvement or cure. Surveys of caregivers have often revealed that the most discouraging aspect is the personality change which estranges the former sympathetic relationship. People say things such as "I feel I'm caring for a total stranger who no longer knows me." Much has been written about the burden of caregivers in the last ten years (5,6,7). They may be physically tired not only from the caregiving and lack of sleep, but also from feelings of loneliness and isolation, especially if they have a very limited social life. Some are physically or verbally abused. People vary in their reactions to these stresses and to their caregiving role. The Zarit Index attempts to measure the burden (8). Not all the effects of caregiving are negative. Some families may draw closer together. Other families may be drawn into conflict. An educated family with good communication and resources may deal with the long term stress best of all. Many caregivers report that they are distressed by the lack of family visits and by the family doubting the illness (9). Middle-aged women in the work force often have a dual role of caring for their parents as well as their own children.

Growth of AD FSGs

The Alzheimer's Disease and Related Disorders Association (ADRDA) was founded in 1980 by several family support groups based in Seattle, Washington (10), Boston, Columbus, Minneapolis, New York City, Pittsburgh and San Francisco (11). With many dedicated lay and professional people, within eight years the organization became 30,000 strong (12).

There are now 188 chapters and 1200 support groups. This phenomenal growth has been fostered by the commitment to caregivers, volunteers and many professional people, some of whom themselves are family members of someone with AD. In the summer of 1988, the ADRDA introduced a new logo and a change in title. It was to be called, "The Alzheimer's Association--Someone to Stand By You" (13). There are also AD associations in 22 other countries which together form the Alzheimer's Disease International (ADI). For example, the Alzheimer Society of Canada was founded in 1978 (1320 Yonge St., Suite 302, Toronto, Ontario, Canada M4T 1X2).

The headquarters of the American organization is in Chicago. The association has, like the federal system, ten geographical regions, each with two elected regional delegates to the National Board of Directors. Region II is composed of New York and New Jersey.

Family Support Groups (FSGs) form the outreach arm of chapters and also the frontline for helping families. They serve neighborhoods and small communities, thus minimizing traveling and absent time from caregiving duties. Common problems such as access and availability of services can be addressed in a locally germane context. Examples of such interchanges include the quality and availability of local physicians and community medical services, home aide and nursing services, the attitudes of local social service offices, and the availability and skill of nearby nursing homes. By inviting professionals from these facilities to make presentations, information flow is maximized in each direction.

The national newsletter is disseminated by all chapters to its members. The Association has also produced more than two dozen brochures.

By 1988, the Association had raised $10 million for research. Through Association efforts the federal government has also increased the amount of money for NIH basic research from $22.3 million in 1983 to $65 million in 1987.

A lobbying firm hired by the Association works in Washington. The Association, with other national organizations such as AARP, is affiliated with the national consortium, "Long Term Care."

The Autopsy Assistance Network was established several years ago to confirm diagnosis and to assist in research by the donation of brain tissue. Families are helped by network representatives to plan ahead and become informed about what to do when the family member dies. With greater interest in the genetic control of inheritance in some forms of the disease, family interest is often stimulated in obtaining an accurate postmortem diagnosis. Local pathologists are encouraged to help in performing the autopsies.

Why Are Groups Needed?

Home caregivers generally do not have an abundance of spare time or money. Groups in the organization are free and open to all, and most communities do not have many professional people for private counseling. It has been said that the real experts are the home caregivers themselves who accumulate so much valuable experience. A group can help caregivers to obtain health care and take better care of themselves. When it comes time to institutionalize a person, it is a very stressful time for both the patient and the family. A nursing home support group can help to make that transition easier for the staff and the family, by explaining the routines, answering questions and encouraging more meaningful visits. The caregiver should also be involved in the settling-in process in the first few days, continuing to help the patient and also with the care plan. This gives the staff more confidence in handling the new arrival and can create a more positive relationship between staff and family in which they can support each other in managing the AD patient. It may also allay some of the guilt that a caregiver feels who is reluctant to "give up." Special groups have been formed for teenagers and grandchildren who live with cognitively impaired adults, and also for offspring and siblings. Personal care aides and aides who work in nursing homes have been encouraged to attend community support groups for their own education in learning about management.

A group can provide support and information and allow people to express feelings and to relate caregiving experiences. A group can become cohesive because of the mutuality its members share with one another. The group can be formal (closed) or informal (open-ended).

The caregiver burden has been discussed extensively, and the need for support is clear. Support can be given in several different ways. It may be respite for the caregiver, practical help in the home, financial support or moral family support, as well as group support. Often a caregiver will use a group initially for basic information in the early stages and then return later when there are more problems. Researchers are trying to evaluate the attendance at support meetings (14).

Facts about the probable course of Alzheimer's disease and other dementias and the importance of early medical assessment to eliminate other curable diseases are often included in group programs. Opportunities to ask questions are usually given to check information.

The daily management of dementia is peculiar to each person, and much of it may be trial and error, but caregivers learn a lot by sharing ideas. Chapters also have books about management available for sale at meetings, such as The 36-Hour Day by

Mace and Rabins (15), The Care of Alzheimer's Patients by Lisa Gwyther (16), and Understanding Alzheimer's Disease, edited by Miriam Aronson (17). Other popular books on dementia include The Loss of Self by Cohen and Eisdorfer (18), and Alzheimer's Disease: A Guide for Families by Powell (19). The caregiver has to learn to be an advocate for the patient to prevent excess disability, by reporting falls and adverse drug reactions, and watching for signs and symptoms of disease such as a high temperature, pain, depression or frequency (20).

People need to know about the home care agencies, mobile geriatric teams, mental health services, adult protective services, Alzheimer's Disease Assistance Centers (ADACs), and other formal support services such as Adult Day Care; area agencies on aging; ombudsman services; congregate meal sites; legal services; respite opportunities, both in and out of the home; local rest homes; and nursing homes. Support group members may demonstrate a need and promote more local resources such as adult day care.

The local chapter distributes the National Association newsletter, which has accurate information about current research and recent findings. Area researchers can also be invited to speak. Chapter membership dues, memorial and gift contributions and local fund-raising activities raise money for chapter education and outreach activities. National solicitation campaigns are also conducted several times a year.

The national newsletter often contains material which relates to state and national legislation, so that people may contact their elected representatives. This is encouraged at the chapter and group grass roots level. For example, in New York a legislative person has been hired at the state level to work in Albany with a representative committee.

The group process may help caregivers to plan for the future, decide which strategies and coping mechanisms work best for them, reduce their burden of caregiving, and learn that they must also care for themselves. Most caregivers feel that they are "not alone" and that their situation is not entirely hopeless through their mutual sharing (or universality) in the group.

Current Research About Groups

There are negative aspects of support groups. People come with different expectations and may leave feeling that they are left out or not helped, and so it is important for the facilitator to be sensitive to those people. Also, people learn to cope with life problems differently and also react to the caregiving role differently, so it is difficult to know which people benefit the most from a support group experience (21,22,23).

I think that only a small percentage of home caregivers attend meetings. There are hindrances, such as distance, inconvenient time, not being able to drive, not finding an adult sitter, not feeling comfortable about going to a strange group, not knowing about the meeting, or feeling too tired or discouraged. A personal invitation is often needed. Visting public health nurses and area agencies on aging may encounter these families and give them this information. One of the research awards given by the National Association this year was to explore hindrances in attendance at meetings. We should continue to reach out in all ways to help the home caregivers and the institutional caregivers of demented people.

Leadership

Leadership of a group may come from a professional or a trained lay person, such as a former caregiver who has time and experience (24). Some professionals who begin self-help groups later take a more advisory role. All business is conducted at the chapter level and there is no hierarchy in a support group meeting. This can be emphasized by informal seating arrangements.

The leader should have some organizational ability for gathering resources and information, inviting professional speakers, putting the meeting together, following up on visitors and keeping in touch with members. Later, most of these things can be done by other members who have time and who, in turn, may be interested in chapter board responsibilities, thus maintaining a "grass roots" commitment to the plight of AD people and their families. If the group serves an area which is some distance from the chapter, it is the facilitator's responsibility also to maintain some personal communication with local resourceful people. For example, it is quite important for physicians who are caring for AD patients to be aware of the amount of physical, emotional and financial burden on the caregiver and family. They can recommend help from support groups, individual counseling, or regional assistance centers for a second opinion. Local lawyers, bankers and clergymen can be given literature to help them to understand family problems. Someone at the local level should be prepared to give talks and information to groups.

Follow-up of visitors to the group is very important. This encourages attendance. I have found that a short monthly newsletter has helped to keep a group together and also to keep in touch with those people who find it difficult to go to meetings. This can also be mailed to local professionals who have shown an interest in the group. If an attendance sheet is kept it is also useful for chapter statistics at the end of the year. Some people benefit by more frequent contact between meetings by sharing telephone numbers. It is helpful as well to set up a telephone "Helpline" for emergencies and difficult situations.

The leaders should maintain good communication with the chapter, which is responsible for maintaining family support groups with supplies and leadership. This is done through the Patient and Family Services Committee where FSG leaders can meet together to discuss mutual problems and concerns. Perhaps a new group is needed, a new meeting time established, or a meeting place has to be changed.

How To Get Started

Explore the need and then always work/check with the local/nearest chapter. If there is none, then in New York, contact the nearest Alzheimer's Disease Assistance Center (ADAC) and the local Area Agency on Aging. Find three or four people to work on a steering committee, including a home caregiver. Plan to continue meeting if attendance at the first one demonstrates a need. Decide on a topic, speaker(s) and structural format. Look for a free meeting place where regular meetings could be held. Caregivers are usually elderly and appreciate well-lit places, plenty of parking, and no steps to climb. Nursing homes and churches, which are accessible to the handicapped, are excellent sites. Plan to have resource materials and a sign-in sheet for follow-up. This is very important. Publicity should be done well in advance to all media, senior service centers, meal sites, doctors' offices, area agencies, AARP groups and churches. For future meetings, devise and maintain a simple method for publicizing the support meetings, which can then be handled from home by a caregiver. The Alzheimer's Association published a very helpful Support Group Manual in 1987 about the formation and nature of new support groups. Another manual was published by the Suncoast Gerontology Center (25).

Meetings usually last about two hours. It is a good idea to strictly observe the time limit for the sake of caregivers whose time away from home may be limited. The first hour may be used for giving information, with a speaker or a film. Topics vary from updates on current research, local community resources, legal and ethical matters, to daily management and problems of nutrition and incontinence. This time should also be used to tell about local chapter activities since some support groups may meet many miles away from the parent chapter. This special relationship can be fostered by leaders who are board members or members of the Patient and Family Services Committee. The chapter's responsibility is to furnish all resource materials and current information about research and policy issues. The information table should always be available with free handouts and books for sale. The second hour can be support time. Groups of about ten or twelve people seem to be a reasonable size for sharing and allowing everyone to take part. Where there is a large number, it may be best to divide into separate circles using each corner of the room with one leader for each group. It is most helpful to have a separate group for newcomers. This provides a special opportunity for them to ask a burning question, to observe the other groups from a distance and makes them feel more comfortable about coming again. Also, new members often have similar questions and concerns. Some chapters ask new members to register ahead of time and have a special closed meeting for them, during which they present a basic information program with lots of time for questions. Two general rules need to be mentioned--to maintain confidentiality and to be non-judgmental. An accepting environment is helpful where people who are full of tension are put at their ease. Refreshments give an opportunity for a social break and allow people to talk about other things as well. Picnics, covered dish suppers and restaurant meals for the group provide some respite from daily care. Sometimes students who are writing papers ask if they may attend meetings, especially for the informational part of the program. It is not helpful to have them recording or writing notes during the support session, unless the group gives them permission. Students and reporters should be introduced and their reason for being there should be explained to the group. They should all be reminded about confidentiality and that permission must be given to quote particular life experiences. Being there is important; even if only one person attends, it may fill a great need. A support group provides a rewarding process to watch the interdependence and mutuality develop toward one another and the acceptance of new members.

REFERENCES

1. Glaze, B. (1982). A Never-ending funeral. *Generations*, 741-52.
2. Barnes, R. F., Raskind, M. A., Scott, M. (1981). Problems of families caring for Alzheimer patients: use of a support group. *Journal of American Geriatric Society, 29*:80-89.
3. Gottlieb, B. H. (1983). Social support as a focus for integrative research in psychology. *American Psychologist*, March:278-287.
4. Cutler, L. (1985). Counseling caregivers. *Generations*, 10:52-57.
5. Gwyther, L. P., Matteson, M. A. (1983). Care for the caregivers. *Journal of Gerontological Nursing, 9*:93-116.
6. Rabins, P., Mace, N. Lucas, M. (1982). The impact of dementia on the family. *Journal of the American Medical Association, 248*:333-335.
7. O'Quin, J. O., McGraw, A. D. (1985). The burdened caregiver: An overview. *Senile Dementia of the Alzheimer Type*. Edited by Hutton, J. T., Kennedy, A. D. New York: Alan R. Liss.
8. Zarit, S. H., Reever, K. E., Bach-Peterson, J. (1980). Relatives of the impaired elderly: Correlates of feelings of burden. *The Gerontologist, 20*:649-655.
9. Scott, J. P., Roberto, K. A., Hutton, J. T. (1985). Family conflicts in caring for the Alzheimer's patient. *Senile Dementia of the Alzheimer Type*, Edited by Hutton, J. T., Kennedy, A. D. New York: Alan R. Liss.
10. Easterly, W. (1982). ASSIST: A model program of support. *Generations*, 7:44-51.
11. Lombardo, N. (1988). *Birth and Evolution in Understanding Alzheimer's Disease*. Edited by Aronson, M. K. New York: Charles Scribner's Sons.
12. Stone, J. (1982). The self-help movement: Forming a national organization. *Generations*, 739-40.
13. Alzheimer's Association Newsletter 8. (1988).
14. Lund, D. A., Caserta, M. S., Wright, S. D. (Nov. 19, 1988). Attendance at support group meetings and burden: A national sample of caregivers to dementia victims. 41st Annual Meeting of Gerontological Society of America.

15. Mace, N. L. and Rabins, P. V. (1981). *The 36-Hour Day.* Baltimore: Johns Hopkins University Press.
16. Gwyther, L. P. (1985). The care of Alzheimer's patients: A manual for nursing home staff. Chicago: Alzheimer's Disease and Related Disorders Association and American Health Care Association.
17. Aronson, M. (1988). *Understanding Alzheimer's Disease: What It Is and How to Cope with it.* New York: Charles Scribner's Sons.
18. Cohen, D., Eisdorfer, D. (1986). *The Loss of Self.* New York: W. W. Norton and Co.
19. Powell, L. S., Courtice, K. (1983). *Alzheimer's Disease: A Guide for Families.* Reading, MA.: Addison-Wesley.
20. CLark, N. M., Rakowski, W. (1983). Family caregivers of older adults: Improving helping skills. *The Gerontologist, 23*:637-642.
21. Haley, W. E., Brown, S. L., Levine, M. A. (1987). Experimental evaluation of the effectiveness of group intervention for dementia caregivers. *The Gerontologist, 27*:376-382.
22. Kahan, J., Kemp, B., Staples, F. R. (1985). Decreasing the burden in families caring for a relative with a dementing illness. *Journal of the American Geriatric Society, 33*:664-670.
23. Wright, S. D., Lund, D. A., Pett, M. A. (1987). The assessment of support group experiences by caregivers of dementia patients. *Clinical Gerontologist, 6*:35-59.
24. Gwyther, L. P. (1983). Caregiver self-help groups: Roles for professionals. *Generations, 7*:37-53.
25. Middleton, L. (1984). Alzheimer's family support groups: A manual for group facilitators. Tampa, Florida: Suncoast Gerontology Center.

Contributors

Howard Bergman, M.D., C.C.F.P., C.S.P.Q.
Assistant Director, Division of Geriatrics
Sir Mortimer B. Davis - Jewish General Hospital
Assistant Professor, Departments of Family Medicine and
Medicine and McGill Centre for Studies in Aging
McGill University
Montreal, Canada

Walter A. Bradley, D.M., F.R.C.P.
Chair, Department of Neurology, College of Medicine
University of Vermont
Burlington, Vermont

Lory E. Bright-Long, M.D.
Health Sciences Center
Assistant Professor of Psychiatry
State University of New York at Stony Brook
Director, Long Island New York Alzheimer's
Disease Assistance Center
Director, Lutheran Center for Aging
Stony Brook, New York

Diane Buchanan, R. N., B.S.
Project Coordinator, Alzheimer's Regional Management Services
Visiting Nurses Association
Troy, New York

A. Mark Clarfield, M.D., C.C.F.P., F.R.C.P.C.
Chief, Division of Geriatrics
Sir Mortimer B. Davis - Jewish General Hospital
Associate Professor and Assistant Dean
McGill University
Montreal, Canada

Raymond A. Domenico, Ph.D.
Chair, Department of Hearing and Speech Science
Co-Director, Northeastern New York Alzheimer's
Disease Assistance Center
State University of New York
Plattsburgh, New York

John A. Edwards, M.D., F.A.C.P., F.R.C.P.
Director of Geriatric Medicine, Buffalo General Hospital
Professor, Departments of Medicine and Family Medicine
State University of New York at Buffalo
Medical Director, Skilled Nursing Facility
Deaconess Center and Episcopal Church Home, Buffalo
Director, Alzheimer's Disease Assistance Center
of Western New York
Buffalo, New York

Richard J. Ham, M.D.
SUNY Distinguished Chair in Geriatric Medicine
Professor of Medicine, State University of New York
Health Science Center at Syracuse
Director, Alzheimer's Disease Assistance Center of Central New York
Syracuse, New York

David Howells, S.C.
Executive Director
Senior Care Connection, Inc.
Troy, New York

Suzanne Lavin, R.N., M.S.
Program Director, Mobile Geriatric Team
Mohawk Valley Psychiatric Center
Vice-President, Mohawk Valley Chapter
Alzheimer's Disease Association
Utica, New York

Robert A. Murden, M.D.
Formerly Co-Director, State University of New York-Brooklyn
Alzheimer's Disease Assistance Center
Assistant Professor of Medicine
Department of Medicine, University of Kansas Medical Center
Kansas City, Kansas

Paul A. Newhouse, M.D.
Director, Geriatric Psychiatry Service
Neuroscience Research Unit, Department of Psychiatry
University of Vermont College of Medicine
Burlington, Vermont

Elizabeth Pohlmann, R.N., M.P.H.
Director, Alzheimer's Disease Assistance Center of the Capital Region
The Eddy Center for Aging
Troy, New York

Alice Barbara Vickers, R.N.
Co-Founder, Capital Region Alzheimer's Disease
and Related Disorders Association
Albany, New York

Taher Zandi, Ph.D.
Associate Professor of Psychology
Director, Northeastern New York Alzheimer's Disease
Assistance Center
State University of New York
Plattsburgh, New York

AUTHOR INDEX

SUBJECT INDEX